Re-engineering Your Business

Daniel Morris

Joel Brandon

McGraw-Hill, Inc.

New York   St. Louis   San Francisco   Auckland   Bogotá
Caracas   Lisbon   London   Madrid   Mexico   Milan
Montreal   New Delhi   Paris   San Juan   São Paulo
Singapore   Sydney   Tokyo   Toronto

Library of Congress Cataloging-in-Publication Data

Morris, Daniel.
    Re-engineering your business / Daniel Morris, Joel Brandon.
        p.   cm.
    Includes index.
    ISBN 0-07-043178-7
    1. Organizational change.   2. Strategic planning.   3. Success in
Business.   I. Brandon, Joel.
HD58.8.M65   1993
658.4'063—dc20   92-38709
                                                               CIP

    2 3 4 5 6 7 8 9 0   DOC/DOC   9 8 7 6 5 4 3

ISBN 0-07-043178-7

*The sponsoring editor for this book was James H. Bessent, Jr., the editing
supervisor was Fred Dahl, and the production supervisor was Suzanne W.
Babeuf. It was set in Baskerville by Inkwell Publishing Services.*

Printed and bound by R. R. Donnelley & Sons Company.

This book is printed on recycled, acid-free paper containing a minimum of 50% recycled de-
inked fiber.

Contents

Preface

Business must constantly improve, and improvement requires change. But how is change to be achieved? There are many and various viewpoints, which encompass little common ground. For example, the executive and line manager have separate perspectives: The manager works within a framework of corporate goals and budgets, while the executive uses these as tools to achieve the business's prosperity. An executive is challenged to do what must be done for the business regardless of the availability of resources, but he or she may—indeed must—make whatever changes in the framework are necessary. Therefore, the two levels have difficulty in cooperating to effect changes, even when the need to change is agreed on. Of course there are also fundamental differences between the viewpoints of managers and the working levels of business.

Philosophies of change management are also diverse. Industrial engineering treats the business like a machine, and approaches change by designing a new mechanical model of the business. Organizational development concerns itself with the psychology of work, and approaches change by motivating the workforce to align itself with the new goals of the business. Quality theorists view business as an entity that reviews the results of its work as it does the work, and feeds the results back into its processes, so as to continually improve. General management approaches change as it would any project, by dividing it into small tasks, assigning the tasks, and tracking their progress on Gantt charts—the "just do it" approach. All four of these approaches are valid. Each has many successes to prove its value. However, it has not seemed possible to combine them effectively. Industrial engineering and organization development seem diametrically opposed, for example, and many very successful general managers feel that the time and effort required for any approach beyond old-fashioned project management is wasted.

However, the changes that business must make are becoming more complex. Standard, time-tested change management methods cannot contend with the new complexities of large business processes supported by rapidly evolving technologies. The more advanced approaches can no longer be looked upon as experimental or as luxuries to be afforded only by industry leaders. They are now necessary to ensure the survival of every business.

This book presents an approach to change management that provides a common basis for the four principal schools of thought. Business process re-engineering, which almost every experienced manager has done in small scale, can be managed in complex situations with full control, using all of what management and management science know about what will work and what will not. Additionally, the support of new technologies can be designed into new business processes in an effective and controlled manner. Our intention is to give enlightened, concerned managers tools to help them achieve what they know to be productive improvements.

Our last book, *Relational Systems Development*, presented many of the basic process design methods that have been refined for use in re-engineering business processes. It was in using these methods in our consulting practice (Morris, Tokarski, Brandon & Co.) that we developed the background that made this book possible. *Relational Systems Development* is itself based on methods that have been used by many management and systems professionals, including various organization and process charting techniques.

Acknowledgments

We had much help in the actual writing: Cynthia Brandon was the principal manuscript editor, Wendy Morris the principal reviewer. Joseph Tokarski, our partner, implemented the methods described in the book in a computer system, called the Positioning and Re-engineering Systems (PARS), which has become our company's primary product. We would also like to thank the professional re-engineering consultants, business managers, and executives, whose ideas have shaped our methods.

Daniel Morris
Joel Brandon

1
Dynamic Business Re-engineering

You can choose to re-engineer, or you can choose to go out of business. Eventually—probably sooner rather than later—every business will need to change the way it conducts its operations. The alternative is to be overpowered by competition. This seems an unpleasant choice, but the nature of business has changed: competition is the most important factor in commerce. Little opportunity remains to do business in a safe niche.

So the question is not whether to change, but how to change. Again, there is very little choice. Cutting cost by cutting budgets and trying to reduce the workforce has been attempted. Introducing quality programs to existing business processes has also been tried. These methods have failed to provide more than very short-term solutions. The new approach is re-engineering: analyzing and altering the basic work processes of the business. Re-engineering is recent, innovative, and not at all commonplace, but it seems to be the only approach that works.

The prospect of re-engineering may actually be very attractive to business people. Re-engineering will allow the full use of their knowledge to be applied to their businesses. In re-engineering, all of what is known to work can be incorporated, and all that is known not to work can be avoided. This is a new opportunity: all business people are aware of needed improvements, and most are eager to apply their ideas.

Little has been written about business re-engineering. Most of the published work has been about particular case studies; there has been almost nothing written about method. This book is intended to be expository and instructive: it presents both an overview of the issues related to re-engineering and methods by which the re-engineering may be accomplished.

The concepts and activities of re-engineering are addressed in depth. The discussion is presented as a "how to" guide that focuses on business processes and how to restructure them. Rather than presenting only case studies, the principles underlying business, work processes, and re-engineering are discussed from the viewpoint of experienced managers and consultants.

The detailed aspects of re-engineering and repositioning are presented in terms of the Dynamic Business Re-engineering methodology, which was developed by the authors in support of our consulting practice. This methodology uses an original, dynamic approach to managing change in business. It does not rely on a static model of the business operation that needs traumatic change from time to time to maintain competitiveness. Dynamic Business Re-engineering is a new approach: it is designed to control change, improve operational responsiveness and quality, and help businesses compete in this new business age.

Who Should Read This Book?

Activities related to re-engineering business processes are becoming more and more important, and their use is rapidly becoming ubiquitous. The term "re-engineering" will soon be used to describe any corporate or institutional reorganization, whether business processes are being changed or not. As such, re-engineering will affect almost everyone in medium to large companies or nonprofit institutions.

This book is written primarily for those who are directly involved in the re-engineering of their companies: all levels of management and many staff members. Re-engineering projects will also be of special interest to the organization's support areas, such as the human resources department, the information systems department, and the financial services groups. Beyond the managers and staff directly identified as part of the project, all of the corporate staff who will be affected should be interested in knowing more about the methods and results of re-engineering.

Although a few managers have titles that identify them as responsible for re-engineering, the efforts will generally be performed by managers and staff responsible for other work. Consequently, the subject matter is presented with no bias toward a particular business area, such as finance or technology.

There is also no bias as to the size of the organization. Re-engineering is most needed in larger organizations that have been in business for some time, but its application can benefit companies of all sizes and the methods are the same. The same is true for all industries, from agriculture to the manufacture of advanced technology products, and for institutions from hospitals to government.

Everyone engaged in any organized endeavor will benefit from knowing about the methods and capabilities of re-engineering, and will eventually be directly affected by them.

What Should You Get Out of This Book?

The subjects covered in this book provide an essential background for participation in re-engineering. However, reading this book will probably not immediately qualify everyone to conduct a full-scale project to transform an old company into a new, globally competitive business. Nevertheless, the preliminary understanding of the process is vital to all re-engineering efforts.

Understand Business Process and Organizational Behavior

The first step toward re-engineering business processes is to know how these processes work and how the organizations that perform them behave. Organizational behavior considerations are important in two ways. First, the exact functions of a business process are best analyzed when the underlying behavior of the groups performing them is understood. Second, new business process designs will be effective only if they take organizational behavior into account.

Organizational behavior is an area of study for which a significant body of knowledge exists. A complete understanding of this behavior is the province of an expert. Indeed, many re-engineering projects should enroll such experts, either from the company's staff (there are a growing number of organization development experts in large companies) or from consulting firms. However, the amount of organization behavior knowledge that is required of the average re-engineering project participant is fortunately modest.

Business processes are the primary target of re-engineering efforts conducted using the approach presented in this book. While this may seem straightforward and possibly simple, it is not. Processes are often not easily defined or understood. And, when examined in detail, most are quite complex. But precisely what are business processes? What do they do? How can they be described? How can they be designed? What are the

trends that business processes exhibit over time? How can costs be associated with them? How is each process related to the others in a company?
Part of the problem is that there are no simple answers to these questions—they change from company to company.

When looking at processes, each level of detail that is uncovered seems
to reveal additional ones beneath it. Knowing how, and why, work is done
by a business at each of these levels is the foundation of re-engineering
and the key to success.

Know How to Position (or Reposition) the Business

Before re-engineering can be done, new goals must be set and a new
foundation laid. We use the term *positioning* (which might also be called
repositioning,) to describe the effort that gathers requirements, set goals,
lays out new infrastructure, and in general repositions the business for
new ways to do its work. Included in positioning is determining the new
role for the company in the marketplace, and planning the steps to get it
there. Another key element in repositioning is determining new corporate strategies and business paradigms that are best suited to the company's new ambitions.

Positioning, as a required preliminary to re-engineering, is one of the
innovations introduced in this book. It is not commonly a part of re-
engineering, but omitting it is a serious defect. With positioning, re-
engineering becomes a much more powerful tool and its effects are more
controllable.

Know How to Re-engineer: A Method That Works

Methodology is another missing link in current re-engineering practice.
Given that redesigning business processes is a good idea, how is it to be
done? The following discussion presents a method, Dynamic Business
Re-engineering, that is based on the Relational Systems Development
(RSD) methods developed by the authors ten years ago to help companies
integrate their business activities and computer support into a cohesive
operation.

Using the methodology for re-engineering presented in this book has
proven capable to providing a systematic, uniform approach to analyzing
and redesigning business processes. Through its techniques it provides

the ability to model proposed designs, observe their full impact on the business, and to compare the costs and benefits of the alternatives.

Because the use of any methodology requires experience and skill, just reading this book will not fully develop the ability to re-engineer. However, it will provide as good a start as can be obtained without actual practice. The book can also be used by those who are engaged in re-engineering projects as a reference and a working checklist.

Know How to Control Change

Another problematic phase of re-engineering is implementation. Often the good intentions of a project are lost due to the inability to translate them into action. The process of controlling the changes required to implement re-engineering is built into the Dynamic Business Re-engineering methodology. It is concerned with not only the procedural changes that must be made to the business process, but also with all the other equally important associated changes.

Know How to Use Change to Obtain Competitive Advantage

The assumption behind business re-engineering is that a streamlined process will produce a lower-cost product (or service) and increase the quality. It is presumed that these factors will make the product more competitive. However, this approach fails to consider such things as customer desires and market pressures. We suggest that a more certain approach would be to first determine what sort of competitive advantage (and how much of it) is best for each situation and then use re-engineering to obtain it.

Furthermore, most re-engineering efforts of the recent past have assumed that the new processes will yield a long-term competitive advantage. They have been "one-shot" efforts. Most have targeted a discrete set of changes that take place over a very short period. It was assumed that these changes will provide as perfect an operation as is possible. However, experience has proven this assumption to be wrong.

To assure competitive advantage over a long term, re-engineering should be done on a repeated basis. The business should position itself to change its processes quickly whenever there is an advantage in doing so. This ability is another of the new approaches described in this book; it is a key element in a company's ability to adapt and take advantage of opportunity.

How Should You Use This Book?

Four general topics are covered in *Re-engineering Your Business*:

1. Chapters 1, 2, and 3 contain background concerning the application of re-engineering to business. These chapters cover business topics related to re-engineering and the uses that have been made of re-engineering methods.
2. Chapter 4 covers the concept of positioning and the management of change in corporations.
3. Chapter 5 discusses the tools used in re-engineering.
4. Chapters 6, 7, 8, and 9 describe how to position and re-engineer businesses, technology, and human resources.

Given today's time constraints, many readers skip around a book to review the topics that they are most interested in. While some may wish to move directly to Chap. 5 and begin with the topics directly related to the Dynamic Business Re-engineering Method, the caution is that the re-engineering presentation may be difficult to understand fully without first reading the initial four chapters. The early chapters should therefore be given as much attention as time permits.

Why Are People Talking About Re-engineering?

Re-engineering is a popular topic in many companies today. Like all new activities, it has been given a wide variety of names, including streamlining, transformation, and restructuring. However, regardless of the name, the goal is almost always the same: increased ability to compete through cost reduction. This objective is constant and applies equally for the production of goods or the delivery of services.

The recent surge of re-engineering efforts is not based on the invention of new management techniques. Industrial engineering, time and motion studies, managerial economics, operations research, and systems analysis have all been concerned with business processes for several decades. The new emphasis is due almost entirely to the recent recognition of an increasing need to compete in order for a business to succeed or even survive.

Market economics is the force that most often motivates re-engineering. Management and engineering methodologies must keep up with the new demands made by the marketplace. Most companies not only recognize this fact, but are taking steps to change the ways of the past and improve in all areas.

Business Is Being Pressured to Change

That business worldwide is undergoing fundamental changes comes as no surprise. But the transition of business is just beginning. Both the character and the extent of the changes are in doubt. The nature of the changes is not well understood, but the increase in competition is clearly apparent.

The most noticeable competition is in consumer goods being produced by Japan, South Korea, Taiwan, and Singapore. However, the real transition is much more widespread. A new industry may be established almost anywhere and compete in a worldwide market. The situation is further complicated by the growth of global assets: companies everywhere investing in foreign ventures of all kinds. Suppliers are also becoming internationalized. The trend toward increases in internationalization will increase competition in the near future even more than it has in the recent past.

The most visible result of these changes is the decline of long established businesses. Some actually have failed completely, and it is probable that more will do so. The loss of market share and income experienced by such companies as General Motors and IBM tend to depress the entire economy. Real estate depreciates in the geographic areas that are most directly influenced by competition, and the financial services sector, in turn, experiences losses. Some of the decline in the world economy identified as a temporary recession is actually caused by a loss of business in the United States.

Business has already realized that increased competition will be the dominating issue of at least the next decade. In response to growing pressures, many businesses have tried to cut costs to keep their product or service costs at a competitive level. These cost-cutting efforts have been generally limited to simple staff reductions and financial manipulations with short-term goals.

The pressure to change is real. It is recognized, and it is taken seriously. However, the response has been limited and not very effective. Most important, long-term planning in response to increased competition is not widely evident. The three areas in which longer views are essential are capital markets, government coordination, and corporate planning. The effect on business of short-term security positions and the conditions under which venture capital is made available currently do not show a concern for developing long-term competitive advantage. The governments of the countries that are most competitive support industrial development. The U.S. government is not among them. Most important, competitive advantage will not come by accident in this new age of business; corporations must plan for it, structure themselves to achieve it, and continue to improve their positions even after they have it. Not one of these three areas has been demonstrably improved in the world's older, traditionally major, economies since the new age of competition began.

But How to Change?

Some progressive companies have seen that their response to the challenge of competition must go beyond just cutting budgets. They have seen that the changes must be effective; not only must costs be lowered but quality must be improved. Selecting specific products or services, they have revised their business processes in various ways to improve their competitive positions. This approach, which has been given the name "business re-engineering," appears to answer the key question of how to change with the new times.

The term "re-engineering" is derived from the practice of information systems development. For some time, perhaps since computers have been used in business, technology professionals have known that the best way to use computers was to enable new, improved business processes, rather than to automate old ones. By fortunate coincidence, the information systems developers were beginning to make progress in implementing business process re-engineering at the time when the need for it became a priority. The application of re-engineering, however, is not necessarily based on the implementation of a new information technology system.

When Chrysler decided to build their experimental automobile, the Viper, as a production car, they took a fresh look at their product development cycle. They wanted to re-engineer it so that less time and expense would be required to bring a new car to market. The success of the Viper project, from a business viewpoint, was much more important in terms of the re-engineered process than in terms of the new product. Clearly, this project was not motivated by information technology. To be sure, information technology was used to support the new process, but the process redesign came first and the technology considerations second.

The term "re-engineering" may be a misnomer. It implies that the business processes were engineered in the first place. However, most business processes are products of a complex series of deliberate decisions and informal evolution. They are not engineered in the sense of a design being created by professionals and the process being built to the design's specifications. Perhaps business engineering is a better phrase, but, of course, it is not in general use.

Industry's experience with projects that re-engineer business processes has been encouraging. Articles in the various business journals are very positive. The reputation of re-engineering is spreading rapidly. However, no stretch of the imagination could call re-engineering a fully developed, common business practice. A manager interested in re-engineering has no body of knowledge to call on. There is no university offering re-engineering programs. Even consultants with re-engineering experience are hard to find. Furthermore, no methodology for re-engineering business processes has gained common acceptance.

The most troublesome aspect of re-engineering in its present stage of development is that, to be effective, it involves a large scope and requires many skills to implement. Business processes cross organizational lines, and changing one process may affect others. Re-engineering requires expertise in personnel work, industrial engineering and economics, marketing, technologies of various sorts and, of course, the specific work being performed. Re-engineering opens a new door, behind which lie many other doors. Few managers have seen what lies behind these doors.

Defining Positioning and Re-engineering

To avoid proliferating new definitions of "re-engineering," it is best to confine the use of the term to the redesign of business work processes and the implementation of the new designs. However, to contain the scope of this definition and yet discuss the other activities required to make re-engineering deliver its principal benefit—competitive advantage—another term must be used. This term is "positioning," or "repositioning." It addresses a higher-level view and set of concerns and uses re-engineering to implement its directives. Positioning determines what should be re-engineered, and initiates the other activities needed to make re-engineering work.

Positioning: Placing Your Company in the Marketplace

Positioning is a set of activities that provides the input and strategic planning framework for re-engineering, and that implements the methods to support rapid and effective change. It begins by gathering data about the company, or institution, and compares where it is today with where it wants to be. This comparison can be done in terms of position in the marketplace or any other appropriate frame of reference. Public sector organizations and private not-for-profit institutions may elect public trust and mission fulfillment as its primary indicators, for example. For profit-making enterprises, analysis of the marketplace and the competition will almost always be the starting point for a repositioning effort. Using the results of this comparison, it is usually possible to set fairly precise goals in the positioning stage.

The second major element of positioning is gathering information about the way that business is conducted. This information provides a framework for change. It defines the relationships among the company's business units and its business processes. It provides a baseline against which future change will be measured, and it supports the analysis of improvements in cost and effectiveness.

The third part of positioning is creating an environment in which change can be implemented quickly, effectively, and without harm to the organization. The rate of change in business is certain to increase. It can no longer be treated as an enemy; it must become an ally.

Re-engineering: Making Changes in the Company

Re-engineering is an approach to planning and controlling change. Business re-engineering means redesigning business processes and then implementing the new processes. If the full measure of repositioning has been done beforehand, re-engineering will have its goals set and its environment prepared.

The separation of the two concepts—positioning and re-engineering—is important for several reasons. First, the scope of positioning is best set very broadly. The entire enterprise, or one independent division are optimal targets. Re-engineering is best limited to a process or group of processes. Second, re-engineering is already beginning to have frightening connotations among the working levels of business. Repositioning specifically includes setting a favorable and trusting environment, as will be shown in Chaps. 4, 6, and 9, which greatly increases the effectiveness of the subsequent re-engineering efforts. Repositioning also encompasses much of the work that must be done by the firm's most senior managers; isolating it obviously conserves their time.

The disciplines of engineering apply to re-engineering projects. They are done in phases: analysis, design, testing and comparison of design alternatives, selection, and implementation. As such, the practice of re-engineering, which has caught the eye of managers for its ability to make businesses more competitive, can also bring a new measure of order to business in general.

Both positioning and re-engineering are flexible enough to be used for either an entire enterprise or a part of one. Because they divide the business into manageable parts, there is no upper limit on the size of the business to which they can be applied. For practical reasons, very small businesses will not want to use formal re-engineering methods, and the smallest unit of a large business that can use them would be a single business process. Any business process will, however, generally cross organizational lines and will typically touch many of the units in a business. This causes the scope of any single-process re-engineering project to be organizationally independent.

The Basis of Successful Re-engineering

We have found that seven capabilities must be part of re-engineering to make it succeed:

1. The ability to conduct re-engineering in accordance with a comprehensive, systematic methodology.
2. Coordinated management of change for all of the affected business functions.
3. The ability to assess, plan, and implement change on a continuing basis.
4. The ability to analyze the full impact of proposed changes.
5. The ability to model and simulate the proposed changes.
6. The ability to use these models on continuing basis.
7. The ability to associate all of the management parameters of the company with each other.

Without all seven of these capabilities, re-engineering becomes difficult to manage and unpredictable, as well as being restricted to delivering only a small fraction of its potential benefits.

The significance of each of these seven success factors is explained in the following paragraphs.

1. Systematic Methodology for Re-engineering. Re-engineering is too important and complex to be done on the back of an envelope. A fully systematic approach to re-designing the business processes should always be used. Furthermore, this methodology should always begin with a detailed mapping of the current business process.

2. Coordinated Management of Change. Business operations must respond to changes initiated by four forces: competition, regulation, technology, and internal improvement. To best react to change, an operation must be flexible and it must be designed for ongoing modification. Re-engineering represents a systematic response to change. If properly used, it becomes a change methodology, a standard approach for modifying operations. As such, it will encompass many components of the business, such as marketing, corporate planning, quality initiatives, human resources, finance, accounting, information technology, and even physical plant. Because of the high degree of interdependence among these activities, a re-engineering project that ignores these areas will probably fail during implementation. For this same reason, the reverse is also possible: an action external to the re-engineering effort can reduce its effectiveness.

The need to coordinate all the factors involved in corporate change is paramount. The most effective approach is to place re-engineering and all other change activities in an overall framework of change management.

3. Continuing Change. Business process re-engineering almost always encounters two very difficult problems. The first results from the sheer size of the projects; they tend to be very large. Management is justifiably intimidated by re-engineering projects that seem to put the whole fate of the company at risk. Also, some projects require so much elapsed time that their effect will not be realized in time to solve the problems at hand. The second difficulty that seems inherent in re-engineering is that the improvements will give competitive advantage for only a short time.

There is a solution to both of these problems. Re-engineering can be done on a continuing basis. Instead of trying to implement a major project that will restructure the entire corporation, a series of smaller projects can alter the business a little at a time. This approach not only reduces the risk and shortens the delay in getting benefits, it enables the company to keep up continually with its competition, government regulation, and the changing business environment.

Another advantage of continual re-engineering is that this approach allows the company's quality program and re-engineering to be completely and effectively integrated. This continuing approach to quality improvement is, in effect, the implementation of W. Edwards Deming's quality concepts. If properly implemented, a re-engineering methodology can greatly improve the effectiveness of quality efforts by helping them look at whole work processes and also to plan and evaluate the impact of improvements.

4. Impact Analysis. Since processes typically cross organizational lines, a re-engineering approach should provide the ability to analyze the impact that changes in any process will have on all organizational units. Also, because processes normally interact with one another, the ability to anticipate the impact of any change on all the associated processes throughout the business is critical. To do this, it is necessary to understand all the relationships among organization, operation, business function, planning, policies, human resources, and information services support. Based on these relationships, any change can be followed through its associations, to determine the full potential impact of a proposed action.

5. Modeling and Simulation. Fundamental to the re-engineering effort is the ability to simulate the changes that are being proposed. This allows the testing and comparison of any number of alternative designs. This ability is based on the use of business process models and some method by which the costs and benefits of each suggested design can be assessed. A computerized modeling system, of course, provides the easiest way to simulate these alternatives.

It would seem rather risky to do re-engineering of processes without any attempt to simulate the results, but it has been tried. In these cases, the business itself becomes the test-bed for the new process, with only limited opportunity to rectify any part of the design that was found to be unsatisfactory.

6. Continuing Use of Designs. The designs drawn for the new business processes should not be used only in the implementation of the new processes and then discarded. Nor should they be stored on a shelf to gather dust and become obsolete. The re-engineering process costs too much; the designs are too valuable.

The obvious use for the re-engineering designs and models is supporting future re-engineering efforts. If a total quality initiative is implemented, the company will need to change its processes on a frequent basis as improvements are implemented. For control, these activities should be performed following re-engineering methods and all documentation should be updated.

A second and less obvious use of the designs is to support daily business operations. The designs contain information that can be useful in making daily operational decisions, in training and in controlling work performance.

7. Association of Corporate Management Parameters.   To begin re-engineering, the project team will require rapid access to all of the information related to the business processes being re-engineered, the company's plans, the current information systems, organization charts, mission statements, and job descriptions, as well as many other details of business administration and work organization. As important as all this data is to the project, the relationships among the data items are equally so. The re-engineering approach, therefore, must have the ability to gather and combine this management information.

A Methodology That Works

Early attempts at re-engineering, both successful and unsuccessful, lacked systematic methodologies. More recently, several methodologies have been suggested for re-engineering phases; some examples are described in the following chapters. The primary purpose of this book is to present systematic methods for the entire spectrum of business change management, from the beginnings of repositioning to post-re-engineering change control. These methods have been developed by the authors to support their consulting practice; they are proven to work effectively.

The overall method is shown in Fig. 1.1, and briefly introduced in the following paragraphs. It begins with the determination of changes that will help gain competitive advantage, and continues through the various activities that lead to real changes in the business. However, the approach presented is not the plan of a single project. It is best itself employed as a standing business process, used as often as is needed, with each major change proposed becoming a project in itself.

Determine Goals and New Market Position

The first step in moving the company to a new market position is determining what that position should be. Factoring marketing into corporate

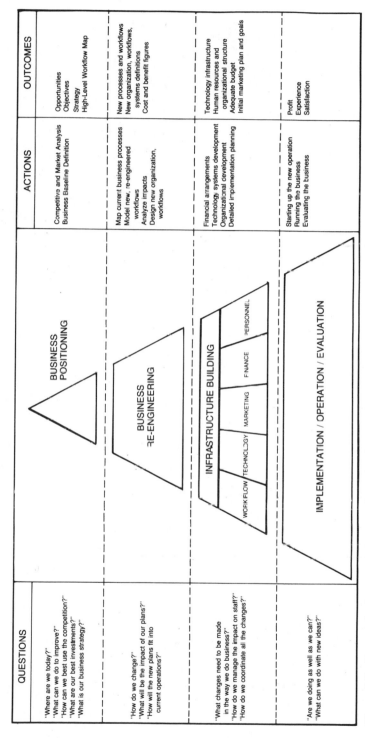

QUESTIONS		ACTIONS	OUTCOMES
"Where are we today?" "What can we do to improve?" "How can we best use the competition?" "What are our best investments?" "What is our business strategy?"	BUSINESS POSITIONING	Competitive and Market Analysis Business Baseline Definition	Opportunities Objectives Strategy High-Level Workflow Map
"How do we change?" "What will be the impact of our plans?" "How will the new plans fit into current operations?"	BUSINESS RE-ENGINEERING	Map current business processes Model new, re-engineered workflows Analyze impacts Design new organization, workflows	New processes and workflows New organization, workflows, systems definitions Cost and benefit figures
"What changes need to be made in the way we do business?" "How do we manage the impact on staff?" "How do we coordinate all the changes?"	INFRASTRUCTURE BUILDING WORK FLOW · TECHNOLOGY · MARKETING · FINANCE · PERSONNEL	Financial arrangements Technology systems development Organizational development Detailed implementation planning	Technology infrastructure Human resources and organizational structure Adequate budget Initial marketing plan and goals
"Are we doing as well as we can?" "What can we do with new ideas?"	IMPLEMENTATION / OPERATION / EVALUATION	Starting up the new operation Running the business Evaluating the business	Profit Experience Satisfaction

Figure 1.1. Dynamic business re-engineering change model.

business plans has always been difficult, but starting with marketing solves many old problems. An analysis of the marketplace comes first. A realistic assessment is made of the company's current position. What will it take to move upward? What about five and ten years in the future? Once an educated business judgment has been made, the corporate goals can be set (perhaps optimistically at this time). These goals should be very ambitious; in today's business world it is usually much better to fall just short of very lofty goals than to meet easy ones and be soundly beaten by the competition.

Establish a New Business Environment

The marketing, expense, quality, pricing, product differentiation, market share, and other business goals that are developed in the positioning process are followed by the establishment of a new business environment. This is formed by examining the conscious and unconscious assumptions that form the paradigms of the company and questioning the ones that are no longer valid. The most important of these is the current attitude toward change and the company's willingness to change when there is a business reason to do so. The new paradigm is one in which change is used to win competitive advantage. As such it must not only be enabled, it must be made entirely acceptable to every member of the corporate staff. The objective of this action is to put the company in a position where change will be implemented without resistance.

Map the Current Business

Once the company has determined its target position and the staff is set for change, the next tasks are to gather information about the company's current business operations and create a model of the business processes being studied. The authors call the basic model a Business Activities Map (BAM). The careful mapping of *current* processes is often given token attention in re-engineering efforts, but, where used, it has been found to be effective and is highly recommended as a starting point for any analysis of business processes.

Business Activity Maps (Fig. 5.3) describe the flow of work in each business process. They are first drawn as very general pictures, showing the main business processes, with no detail and only a few notes related to quantitative information. For example, the average length of time that a process requires may be included. The general, high-level maps are then broken down into more detailed ones, until a very fine level of detail is obtained for all tasks in the process. At detailed levels, more numerical data is obtained and included. The process of adding levels of detail, called "leveling" or "factoring," is not difficult, but it does require the involvement of the staff who actually perform the work. Additional tables, or matrices, are used to capture supporting information for the processes, such as the answers to the basic questions who, what, why, when, how, and where? Facts concerning the use of information in each business process are also sought, as is the usual organization chart data and any existing business plans.

Redesign: Using the Map as a Model

The reward for gathering the details of the current system comes in the next activity: creating a new business process design. The redesign effort uses the BAMs to model the new business operation work flow. The redesign process becomes a modeling activity, with guesswork reduced to a minimum. The first model can be a real attempt at designing the final process, or a "straw man" can be used, in which all similar work activities are combined. This can be called a least-cost design, in which all duplication of effort and redundancy is removed. The least-cost scenario is seldom feasible, however, because other desirable characteristics are sacrificed. In most cases, the managers reviewing the design will create some duplication in successive new versions of the design.

Each version of the model is reviewed by managers and staff to ensure that the work functions are adequately facilitated. The versions are also subjected to cost and timing analysis, so that the attainment of quantified goals can be tested. Using paper or, better still, computerized models to simulate a new operation serves business far better than using trial and error on the organization itself. Usually, the actual implementation is required to be nearly perfect, as most companies will tolerate only a minor fine tuning after a reorganization. However, without modeling, the probability of any reorganization being fully successful is small. This is borne out by industrial experience with reorganizations motivated by cost cutting in recent years.

When the modeling process has produced a satisfactory design, the result will be a map of the new business operation in the same format as the map of the current business. It will be understandable, because it is written in terms of the way things are done. The project will have produced business process designs that are likely to work very well, that have been reviewed by the managers who will work within them, and that can be easily implemented.

Build the Foundation

The new processes will need more than a new organization chart and work flow diagram. The design of the new business is examined by experts in training, staffing, information technology, marketing, accounting, and finance to determine the needs for new support from those areas. These support elements are usually put into service long after the business process has been operational, and they are seldom coordinated. The processes of repositioning and re-engineering provide an opportunity to do better. At this time the impact of support can be assessed, often quantitatively, planned and fully coordinated. Equally important, the support elements can be helpful in smoothly implementing the newly designed processes.

The third level of the Change Model pyramid in Fig. 1.1 represents a level of planning in which the support infrastructure and the architecture of the business are designed. The support infrastructure is composed of the activities that support a business process without being a direct part of it: for example, human resources, travel, and procurement are all support activities. The business architecture is the overall design of the processes and support activities that work together to form the entire enterprise or a large part of it, such as a division. Like an architectural design, the structure of the business is the basis of the architecture, and the support elements are each deigned to have a specific role.

The design of the infrastructure includes consideration for information use, whether or not any technology is required. Information systems and technology are given a high priority in this phase of re-engineering, due to the long lead times and possibly high costs involved. In most medium-sized and large businesses, information technology is a basic requirement in many work processes, and it will grow in importance as competitive advantage is increasingly based on the creative use of information. As new systems will require fairly long lead times for implementation, and new equipment will require some time to obtain and set up, this support element may determine the overall implementation schedule for the re-engineering project.

The next support issue to be adddressed is the human capital required for the new business processes. The infrastructure related to human capital may be built by specialists in organization development, staffing, training, and other human resources topics. It consists of position descriptions, training, and staffing requirements, and various methods to achieve teamwork, quality, and focus. One very important human capital consideration is the method by which the new business processes will be staffed from the old ones. The objective is to avoid the usual personnel problems that seem to be an unavoidable feature of reorganizations. Business re-engineering can help achieve this objective.

Marketing, which contributed heavily at the beginning of the re-engineering project, may re-enter it at this point. Marketing may be linked to a business process to provide support to both the operating and marketing efforts. For example, if order processing is being re-engineered, Marketing may be able to help define address requirements and may want to know as soon as possible when standing orders are canceled. These relationships may or may not be important enough to be noted explicitly in the Business Activity Map models. When such marketing linkages are important to the business, their presence in the Business Activity Maps will assure that they will be given consideration when the new infrastructure is being designed.

Obviously, infrastructure includes such common support as facilities, power, heating, ventilating, air conditioning, telephones, lighting, and office equipment. Given a fully detailed map of the new business processes, these elements can be installed, renovated, converted, or just reassigned with great accuracy. In some cases, however, specialized equipment or facilities will be required and special considerations must be made for putting them into production and for phaseover from the existing equipment.

Finance and Accounting should not be forgotten when building infrastructure. Accounting, in particular, can provide much more direct support to the business operations than is customary in most companies. This phase of re-engineering presents an opportunity to establish meaningful cost accounting for the new processes and to build a management budget structure that directly helps managers.

Implement and Operate

Implementation planning, implementation, and the operation of the new business processes comprise the last level of the Change Model in Fig. 1.1. This last step is where the payoff occurs.

Generally, implementation of a reorganized or restructured business operation is difficult. If there is the slightest hint that management and staff workforce reductions will be a part of the effort, it can be a nightmare—undoubtedly the greatest challenge that a manager can face. The methods that we suggest can help considerably, however. First, management will have comprehensive before and after pictures of the business to work with: the maps of the current and the new, re-engineered business operations, plus the organization charts and other supporting documents developed in earlier phases. Having these tools will place management in much greater control than when organization charts, and perhaps mission statements, were relied on.

Another important aid to phaseover is the prearrangement of infrastructure. Having support already determined avoids scrambling under the pressure of production requirements at the time of phaseover, or even after the fact during operations, in full view of customers. Prearrangement essentially places all of the company's capabilities in support of the implementation, instead of their seeming to be opposed to it.

Operations after the phaseover is completed will continue to be affected by the re-engineering effort. The measurements shown in the Business Activity Map for the new process become the production goals for the new process. The efficiency and performance of the process and its staff can be measured using these numbers and the framework of the Business Activity Map, which clearly shows what is supposed to happen. The re-engineering documentation is also used to support ongoing improvement by assisting all managers and the production workforce in the continual redefinition and attainment of quality in every detail of the work being done.

The New Environment

The final, but perhaps most important, contribution to be made by the methodology presented in this book is the introduction of a new business environment, a new paradigm for change.

This change paradigm is based on the continuing application of repositioning and re-engineering. Currently, re-engineering is generally viewed as a necessary but difficult cure for problems that have grown over time and must be solved, once and for all time. After re-engineering, a new business process should emerge that should beat the competition for a considerable time. This is a shortsighted view. Re-engineering should be done as often as necessary to obtain competitive advantage. When new products are to be produced, when new technologies can reduce costs, when new markets can be opened, and in fact when any significant opportunity appears, the repositioning and re-engineering cycles can be initiated. Another important benefit is the ability to break very large projects into manageable smaller ones. This helps to avoid the long delays associated with large projects and reduces risk. The character of process changes can be altered: instead of being do-or-die attempts to achieve perfection, they can become a succession of controlled improvements.

Frequent, perhaps continual, re-engineering is not as difficult as might be supposed. The two highest hurdles in the methodology are the mapping of the current business and solving the personnel problems associated with restructuring. If these two things are done well during the first re-engineering effort, they will not need to be repeated. The new business process designs will become the current Business Action Maps as soon as the new processes are in production. The staffing problems can be handled in such as manner as to establish a permanent relationship of trust with all staff. Change will no longer be difficult or threatening. It will become the means of success.

2
Understanding Business Behavior

Before delving into the mysteries and complexities of repositioning and re-engineering, the environment in which they are to be used must be examined. This environment is, of course, the business world. How does business behave, and how do businesses behave, with respect to change? Also, what elements of organizational behavior are related to re-engineering? These and other related questions are addressed in this chapter.

How Business Evolved

How did business get where it is? Business has been slowly evolving. The rate of this evolution has, however, dramatically increased in the recent past. Today, the current rate of change in business is high: one of the few fundamental changes in the history of business is happening now. If the nature of this change is understood, it will provide insight into how to use change in business during this period of transformation.

The Common Evolutionary Pattern of Businesses

Business began before recorded history. Archeology has found evidence of commerce in the oldest cities. One theory holds that it began when surpluses in farming first occurred, that the first commerce was trading excess food for other things, and that the deliberate production of eco-

nomic goods followed. Trade may have actually motivated the construction of the first cities; it certainly flourished in them.

If the theory of surpluses is correct, the first motive of business was the
exchange of worthless excess goods for useful ones. The first significant
change would have then been the invention of trading for profit. This was
the beginning of business as we know it today: profit is still our motive,
and it is still earned by selling something for more than its cost. Many of
the other aspects of business also remain as they were then: business
people from the dawn of civilization used money and kept journals of
their transactions.

Today's business is built up from the technologies of the distant past
with only a few major changes over the millennia. The Egyptians invented
paper in about the third millennium B.C.; the next milestone is the publica-
tion of *De Computis et Scriptis*, written by the Franciscan monk Pacioli in
Italy in 1494. In this work, which is a part of a book entitled *Summa* (the
first printed book on the subject of algebra), double entry bookkeeping is
described in almost the exact form in which it is used today. At that time
chemistry, astronomy, medicine, and biology were primitive, but bookkeeping was fully developed.

A few centuries before Pacioli, the city-state of Venice became dominant in world trade, setting new patterns for cooperation between government and business. For example, Venice invented what we now consider normal diplomacy, with the first use of resident ambassadors, along
with espionage and secret codes for diplomatic messages. All of this was in
support of commerce, which was the central concern of Venice, a society
ruled directly by business interests.

The technology of business was based on ledgers of transactions until
the 1870s, when the U.S. Surgeon General's office invented the file folder
and filing cabinets. This was a remarkable precursor to the computer,
which was the very next invention to alter business. Computing in business was initiated by Herman Hollerith, who designed the punched card
to help count the 1890 census for the U.S. Census Bureau. The U.S.
census of 1880 took seven and one half years to count: Hollerith's invention allowed the 1890 census to be counted in only two and a half years,
and the accuracy was much improved. Hollerith left the Census Bureau in
1896 to form the Tabulating Machine Company, which subsequently
became part of IBM. The business accounting machine became the basic
tool for managing the large amounts of data that increasingly large
industries and financial services companies required to do business. This
function was inherited by the stored program digital computer in the
1950's. The personal, desktop computer entered the business world only
a decade ago. Clearly, only a handful of real technical changes occurred in

the first four millennia of business. Most of them have happened in the last hundred years.

Changes in business, most of which are based on new technology, seem to be occurring at an accelerated pace. From the middle ages to the beginning of the century, there was almost no change at all. Accounting did not change from Pacioli's time until very recently. It was therefore understandable and predictable that business would have some difficulty recognizing that fundamental changes were taking place, and that business would also have some initial difficulty in finding the best ways to react.

The way business is conducted may be changed as much by the most recent innovations, such as word processing and electronic mail, in the next few years as it has in the past five thousand years. To survive, it is necessary to find ways to anticipate these changes. It is also reasonable to find ways to profit from them.

How a Business Evolves

Just as business has evolved, each business organization typically experiences a sort of evolution from its incorporation through growth and maturity. When a business starts, it is usually small, and it is operated by a few people who know one another. The first few people in a new business are highly motivated by their close relationships and the knowledge that the success of the business depends upon each of them doing more than a day's work each day. Furthermore, the rewards for individual workers in a new, small company can be very great. Processes, policies, and methods are informal, except for the few required by statutes related to corporate conduct and record keeping. The size and nature of a small business keep business processes in particular simple, straightforward and known by all.

By the time the company has grown to medium size, these informal approaches become obviously insufficient and more formal ones are put in place. At this time, and it can come very quickly in some industries, there is some conflict between the way things have been done and the way the proposed new rules are written. Indeed the first attempts to impose formal organization tend to be overdone and are often undertaken with little appreciation of the corporate culture (see Chap. 4).

When the new corporation gets past this difficult adolescence, it becomes a young adult company and begins its transition from the control of its founders to new management. This can be accelerated by acquisition, but it will come sooner or later, and when it does the business usually changes its rules again. These changes in corporate maturity are usually made with less planning than would be ideal, because either the changes or their magnitude are not anticipated. Underplanned, the changed rules are enforced to some extent by reaction, leading to increases in emergency repairs to organizational structure and process design.

When the corporation reaches maturity, there is the illusion of an ordered existence. There are job descriptions, in theory for every position, there is a uniform compensation policy, more or less, and a performance review system, which is given annual attention only. The average mature corporation has two sets of rules: the formal ones and the ones that are actually followed to get work done. Many have an informal organization chart working in parallel to the formal one as well. There is much bureaucracy, and top management does not control the detailed activities of the working level of the business. The whole organization works, however, because the entire workforce is genuinely concerned that it must. There is a very heavy reliance on the commitment of the workforce in established businesses. Changing things is difficult and often unfair, due in part to the reliance on informal structure to accomplish the work, while using the formal structure as the basis for change.

Established concerns cannot grow forever. Sooner or later, they must experience either some stagnation or some loss, requiring changes such as reductions in workforce. When this occurs, and is complicated by the problems that the corporation has collected in its evolution, the trust, dedication, and motivation of the workforce are eroded. This leads to further problems in operating and in making further changes. It can easily become a descending spiral.

With this evolution behind them, it is not surprising that many of the largest and oldest companies do not welcome change. These companies, however, are the ones that need to change most urgently.

The Evolution of Informal Business Structures

When the business has reached medium size, bureaucracy first begins. The organization and territorial barriers cause communication and work flow problems. Change becomes sluggish and help among business units is restricted by red tape and growing self-interest. Each new change to the business is implemented with the best intentions, and almost all have genuine benefit, but they are applied as additions to the business structure, further complicating it. In time, barriers to progress and creativity will abound.

The managers and workforce of the business will, of course, do their best to work within the boundaries of their business structure, despite its complexity. However, they will also try to improve their individual situations. This can have several effects. One is the creation of informal networks of cooperation, having implicit organization charts and work procedures. Another common effect is the formation of political alliances, which are networks intended to further the selfish interests of the participants above the interests of the business.

Individual managers also attempt to work around the bureaucracy. Out of frustration, they often further narrow the scope of their activity to a point where they can control all the variables. Some managers react by building their organizations rather than narrowing their missions. Some build complex informal networks to support their work. Some others combine all three of these maneuvers in creative ways. For example, these managers will often include management from several levels above their own in their informal networks. The people who use these maneuvers are usually regarded as the best managers in the company. They are admired for their creativity in overcoming the obstacles of the corporate bureaucracy. Unfortunately, by refusing to accept the limitations of the organization, they consistently accomplish their work objectives.

These super achievers do get the job done. However, by further narrowing the scope of the activity, devising shadow organizations, ignoring uncontrollable variables, and creating informal processes, they further weaken an already complex and sluggish business structure. As this happens in many areas in the company, the integrity of the intended business processes are further diluted.

When changes are attempted in business, the informal structures and the wake of the fast-swimming super achiever make analysis of the current business more difficult and also interfere with implementing new organizations and processes. Re-engineering is often confronted with these problems and must be specifically prepared to address them.

The Evolution of Hierarchical Structures

The basic organization plan used by most business is the hierarchy. It starts at the top with the chief executive officer and spreads out until the actual working level is reached. The behavior of these structures over periods of time has subtle nuances that are not well known to business, although the results are. These behavioral patterns have importance to re-engineering efforts.

The first problem becomes evident when designing a hierarchical organization chart. There is only one orientation possible for a hierarchy, which is found to be limiting in an increasing number of cases. For example, a business that organizes itself by product will not necessarily be organized by function, so functions will have to be duplicated for each product or shared somehow. Sharing, however, is not one of the flexibilities of the hierarchical structure. Many companies are now struggling with ways to be organized by, or at least to focus on, all of the following at the same time: *function*, for efficiency's sake; *product*, for the sake of effectiveness; and *client*, to improve customer responsiveness. The hierarchical structure does not support their efforts very well; no hierarchical structure they can devise can focus on more than one of these, forcing them to choose among the three.

The most obvious characteristic of the evolution of hierarchies is the creation of additional layers of management to accommodate organizational growth. This tends to add overhead, or at least reduce the economy of scale that is expected when corporate growth occurs; the need to add staff for a new function should be less than if the new function was to be staffed as a entirely new company. Additional layers of management present more difficulty than adding highly paid staff, although that is bad itself. Added layers generally reduce the ability of the organization to respond. The distance between the lowest level and the highest will add time and effort to decisions in most organizations.

Hierarchies tend to grow in size. When there is a new piece of work to be done, the tendency is to add a whole position for it. This is due to several characteristics of the hierarchical structure. First, each of the lowest positions tends to be a specialty, making adding work to any one position difficult unless it fits into an existing position perfectly. Second, there is no formal relationship among any peers; so redistributing work is the job of the manager. Third, status is given by level; so the managers are motivated to increase the structure beneath their own positions. Moving positions can be difficult, leaving gaps and changing what seem to be balanced conditions. Finally, the compensation structure associated with hierarchies is to pay each position, not to require sharing of revenue; so the more positions, the easier the work, with no reduction in compensation. Thus the hierarchy tends to grow, given the slightest reasons for doing so.

For many of the same reasons, it is difficult to prune positions from a hierarchy. Certainly, the remaining staff have no reason to be happy about picking up additional work. Most managers are aware of the hierarchy's stubborn reluctance to yield positions—even after an efficiency gain has reduced the overall work of the structure. For example, if all of the secretarial work in a company can be reduced by 50 percent, every manager in the structure will react by wanting to keep his or her own secretary, which means that no reduction can be realized. Only a fundamental change in structure will permit any cost savings in these situations.

One other undesirable feature of the evolution of hierarchies is that organizational units tend to become isolated within them. As hierarchies grow, the lack of peer relationships motivate managers to obtain all the resources they need within their own organizations. Clearly, most managers do not want to become dependent on other managers for vital support. As a result, the more the managers in a hierarchy succeed in gaining independence, the more they drift apart. Isolation thus grows, while communication and cooperation decline.

Experience has shown all these mechanisms to be at work as hierarchies evolve. The result is that hierarchies of business have shown tendencies to grow and to become difficult to manage. Isolation of functions is another characteristic of hierarchies in most established businesses, with little detailed knowledge of operations going beyond the immediate organization performing the operation.

Experiments in organizational forms other than hierarchies have been carried out for several decades, beginning with matrix management. A few years ago, a trend toward flattening organization structures began. These changes increased span of control and decreased management layers, with mixed results, but generally with some reduction in management costs. More recently, different types of teamwork are being tried with some success. High-performing teams and self-managing teams, such as D. Quinn Mills's Clusters, are usually set up to share revenues and work more effectively than hierarchies. They are not motivated to grow, and they require less management. These empowered teams, which are discussed in Chaps. 4 and 9, provide a good strategic staffing approach to implementing re-engineering changes.

New Pressures on Business

The changes in the business world since 1970 seem dramatic, comparable in scope and magnitude to the industrial revolution or the beginning of the computer age. There is a shift toward services, the end of the commercial dominance by the United States, a shift toward information uses, globalization of most forms of commerce, and, of course, a very great increase in competition.

As is shown in Fig. 2.1, the pressures for change seem to impinge on every business with cumulative impact. A small amount of pressure from

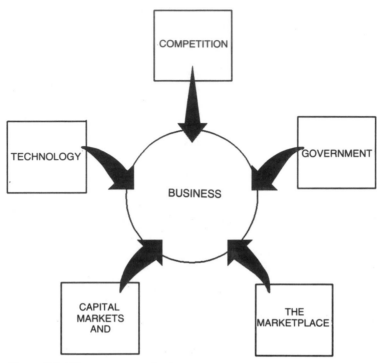

Figure 2.1. The pressures on business.

each of the many change factors may result in a significant total effect. When the pressures line up together in some consistent way, they cause a permanent change in the way business is done. Such changes have occurred many times in history. For example, the leading influence in international trade was once exercised by Venice; subsequently, the British Empire became dominant. When these historical changes began, they were not immediately recognized. Today we are better able to analyze changes in commerce quickly; for example, we know that a major change is taking place now. However, we are still not at all sure of the exact nature of the change and even less sure of the outcome.

An educated guess as to the new, emerging character of business is that it will be in flux, perhaps for a long time. There will be a high level of international competition and no one country will be able to dominate as the British Empire and the United States have done. No market will be dominated by one corporation as IBM has done. Companies will be able to enter the marketplace and succeed with novel products or services or in specialized niches, for a time, but sustained growth will be very difficult to achieve. Information will be a base resource. Societies will become information-based, with an increasing amount of work effort devoted to it, just as the movement to industrial-based societies moved effort from agriculture to industry. All this will require that business be very flexible, and that business people become aware of the international trends in their business areas.

Global Competition

In market economies, competition is considered the most important factor in the business climate. Michael E. Porter has studied competition and shown, in *Competitive Advantage: Creating and Sustaining Superior Performance* (New York: The Free Press, 1985), among his other works, that it is influenced by five factors:

1. The ability of new companies to enter the market.
2. The ability of suppliers to exert cost pressure on the competitors in the market.
3. The ability of customers to influence the competitors (for example, if they are price-sensitive, customers will force price competition).
4. The ability of alternatives to pressure the market (for instance, the use of PCs pressures the computer mainframe market).
5. The competitive activities of the competing companies themselves.

These factors are themselves influenced by other factors, such as government, technology, and globalization, to produce changes in the business environment in individual industries, whole economic sectors, and whole economics. Although such changes have always taken place, the rate of change seems to be accelerating, and the character of change seems to be particularly difficult to manage, mostly due to the new challenges offered global competition.

The globalization of competition is very well known. The overall increase in exporting (for the whole world) between 1985 and 1990 is over US $1533 million, which is an increase of 86 percent.* This large number represents far more than just an increase in Japanese goods being sold in the United States. There is now more foreign trade among all of the world's nations. There is increased foreign investment as well. Why is this happening?

*SOURCE: The International Monetary Fund.

Although a number of reasons have been offered for globalization, it might be more reasonable to ask why it did not happen earlier. Globalization was the policy of the United Nations, which began as a wartime alliance formed to oppose the Axis powers in World War II. The formation of the World Bank and International Monetary Fund shortly after the war was the first concrete expression of this policy, which has continued to the present day. Global trade and the development of new national economies was the primary weapon of the Western powers in the cold war. It has succeeded. Almost all the changes in international business development have been a consequence of this successful policy.

The single most important reason for the success of international businesses is the accuracy with which they have assessed the marketplace. Most of the businesses effective in foreign markets selected their markets, studied them, and then did what was necessary to outdo the standing competition. In many cases they were also starting from no base, which is advantageous in that it is easier to build a new organization than to change an old one, and there were no old investments to protect.

Much has been made of cheap labor as a cause for competitive advantage. However, low-cost labor does not seem to have provided competitive advantage, so much as it has increased globalization of production. Industrial countries with higher wage scales tend to obtain some labor from less developed countries in one way or another. For example, some United States manufacturers will have parts made in Mexico and China, but the overall product is labeled as being manufactured in the United States. Some cases are even more convoluted, but the overall effect is that most manufacturers can avail themselves of cheaper labor, if they are willing to pay more for shipping and have less control over their inventory levels.

Another aspect of the growth of global trade is that many national governments exert considerable pressure on their businesses to export. Export provides an influx of hard currencies and credits, which is an important factor in developing countries whose credit is limited. The pressure exerted by governments takes many forms, including subsidies, capital investment, and tax incentives. Many governments also maintain agencies that assist foreign investors. Foreign industries are invited to buy real estate and set up facilities, again to improve the flow of currency. Even a modest result from either sort of government intervention can greatly improve a national economy. With over 200 countries engaged in these practices, the effect on globalization is significant.

A factor commonly found in countries that have become successful exporters is the use of advanced cooperative relationships among the organizations involved. Industrial management, labor, government, together in new ways, and succeed as a result. These relationships do not remove the regulatory role of government, nor do they place labor and management in complete accord, but they obtain sufficient advantage to place themselves ahead of other countries.

The relationship between industry and government, for example, is cooperative to the extent that national goals are set for key industries, such as the national computer policy of Singapore. Singapore has created a National Computer Board to support the growth of the information technology industries there. The Board sets standards, certifies computer professionals, and helps the educational system to produce enough computer professionals for anticipated needs. The government of Singapore has also set up a communications network, called TradeNet, to connect all of the parties engaged in export, which has cut the time to export goods from many days to a few hours. Japan is also famous for government cooperating with industry, as is the case in South Korea, Taiwan, and Thailand. Indeed it seems that every country that has improved its balance of trade dramatically in recent years has shown this characteristic. An example is the comparison published by the Organization for Economic Cooperation and Development (OECD) in Paris, which shows that, between 1975 and 1985, every country that had implemented a national computing policy experienced an increased balance of trade in electronics, and every country without one experienced a decrease. The relationship between government and industry is apparently more important than is commonly recognized, even with the subject before the public, as it has been recently.

The relationships between capital markets and businesses are also inclined to be more productive in those countries that have shown high economic growth rates. The opinions of securities analysts, and to be fair, the sensitivity of businesses to them, have influenced far too much short-term and financially motivated planning. Banks, too, have been more interested in balance sheets than business plans. It is difficult to change these postures without changing the attitudes of many people. However, there are some lessons from countries with different attitudes. It seems that the ability to take a long-term view (1) is based on a certain confidence in the future, (2) results from direction from the very top of the company (or bank), and (3) can be influenced by government action. This last factor, government's ability to influence a longer planning horizon, comes from setting long-term national plans that raise confidence and engender a national sense of purpose. The government can also influence the orientation of capital markets by using incentives for long-term investments.

International success has also benefitted from another advanced relationship: the one between suppliers and their industrial customers. In the extreme example of *keiretsu kaisha* and *kankei kaisha* (affiliated and associated companies that are essentially a part of a large corporate group) in Japan, the relationship is almost a partnership, and is often viewed as a closed, clannish system, maintained to exclude outsiders, but in fact it is not. In these relationships, the suppliers believe that they will not succeed unless their customers do. The companies that they supply believe that they will get better supplies and lower prices by not using competitive procurement. These attitudes prevail in Japan, replacing the desire of suppliers to get the best prices and terms that they can, and thus forcing their customers' costs upward. The Japanese attitudes lead to very highly tailored, high-quality supplies and just-in-time cooperative inventory management, among other things. Because other elements of supply provision may receive consideration, it may seem that the advantages of market price competition are lost, but this does not necessarily occur.

Political restructuring—the conversion of socialist, government-controlled, systems to market economies—began in 1989 and will have a further impact on globalization in the future. The progress of even the earliest experiments in restructuring in Hungary is not advanced enough to have made a significant impact on world trade. Although political restructuring will increase global competition and the effect will be spread over as much as two decades, it is essentially a single event, incapable of being copied elsewhere, and not a competitive methodology.

Given these methods, competition (global, national and local) has increased and will continue to increase. In general, competitive advantage can be maintained only by making continual improvements in product, cost control, and marketing. This would seem to indicate that increases in competition will continue and even increase in intensity. Clearly, these factors will become the daily concern of management in those companies that learn to survive.

The Marketplace

In addition to the pressures associated with increased global competition, changes in the marketplace are driven by several other sources. These include changes in the buying habits of customers, changes in product life cycles, changes in inventory management and increased demand for quality. These changes have resulted in a demand for increased management intensity and faster response to change. Unfortunately, the overall effect has been a reduction in corporate performance in many industries.

Customers, both individual consumers and corporate buyers, are increasing their demands as they see competition increasing. Brand loyalties seem to be less important. There is less tolerance of poor quality, a predictable consequence of increasing product quality. Today it is almost impossible to successfully market any goods or services that are not basically of good quality.

Fine tuning of management practices is leading to the reduction of oversupply in goods inventories and even in services. This reduction in overstocking and in supplying extra services causes a rippling in the market as the demand from those companies that manage themselves more closely decreases.

The other major market factor is the reduction in the length of time that a product and its associated marketing can remain unchanged and yet profitable. The time required to develop and market new products is being reduced, deliberately in most cases. This makes old products lose market share more quickly. The entire life cycle, from concept through marketing to obsolescence, is much shorter today for many products. This increases the pressures to drop, update, or reposition products and services on a more frequent basis. For each product entering the market, the shorter development time reduces costs and increases revenues. But if the practice becomes widespread, competition is increased and the overall performance of the industries in which it is practiced goes down.

The Public Sector

Changes in the many governmental and public institutions have also contributed to the pressures on business. The influence of taxes and tariffs is known, but it can be subtle in some cases. Monetary policy is also a government issue of immediate importance to business. Regulation is another. Not as well recognized is the effect on the supply of human capital that the government and educational systems have, and its impact on business. The other factor that is overlooked is the effect on business of the government's role as a consumer.

Tax and monetary policy can be viewed as "control knobs" that the government can turn to change the economy. Unfortunately, the proper setting of these dials seems to be elusive. Each control has both positive and negative responses, of course. Setting high interests rates, for example, is believed to reduce inflation, but it also slows the economy. Sometimes the controls fail to work, as the U.S. government found when it lowered interest rates and did not see the great upsurge in the economy that it expected. Because most of the decisions made in setting governmental economic controls is political, there is no guarantee that an optimal climate will even be sought.

Government control has become more intrusive toward business recently due to globalization. This involvement in internationalized business is a result of the need to deal with an increasing number of governments. Still more change has resulted from the reaction of governments to the erosion of their tax bases, which is one of the by-products of greater importing and productivity. The consequences of the increased taxation that often accompanies this erosion are almost all negative. Eventually, the power to tax is the power to destroy, and some of this power is acting against business in the most well established national economies.

Regulation is the second direct government influence. Regulation, policy, and legal framework influence business in ways that are difficult to anticipate or even respond to. The judicial trends toward product liability in the United States are a painful example. Even efforts intended to help business may introduce change that is difficult to manage. For example, the government debates between regulation and deregulation, and the actions ensuing from them, have provided great change to business in recent years. Deregulation and privatization, generally presented as being in the interest of business, can also have very negative effects, resulting in vastly increased competition and a consequent loss of performance for many companies. Certainly major changes in regulatory policy require business to react, and there have been many such changes throughout the world in the last two decades.

Other interesting change factors are the supply, or lack of supply, of human capital and the quality of education. Governments influence all levels of education. Because this is another politically sensitive area, the results of the educational systems are not always going to be what is needed by business. Furthermore, the differential in the quality of graduates among countries may shift, supporting the movement of competitive advantage.

The final and most underrated influence exerted by government arises from its role as a consumer. The public sector taken as a whole is an extremely large consumer, with the largest consumer of all being the U.S. government. Due to this volume, the buying habits of government can influence the competitive environment in any industry. In some areas, government purchasing dominates the marketplace. In these areas, changes in government procurement policies will motivate many businesses to change their product strategies. An example is the adoption of the GOSIP (Government Open Systems Interface Profile) standard for information systems interconnection. GOSIP is not a national standard; it applies only to products purchased by the government. However, every manufacturer of information technology equipment has responded to it.

Technology

Technology is the most important single factor of change. That the rate of technological change has been accelerated in recent years is known to everyone; it is obvious from the stream of new products that have been marketed. The changes that business must make in reaction to technical change go beyond new product designs, however. Much of the influence that technology exerts is upon the way that business is conducted. If technology can either improve quality or reduce costs in any way, it can be used to create an advantage. As a result, all competitive businesses are virtually forced to make use of any new technology that offers significant improvement.

Technological changes are particularly difficult for business because, by their nature, they will always be new and therefore poorly understood. Business technology is evolving so rapidly that one technology is replaced by another long before many businesses have learned about the first one. To compensate, a business is forced to control new technology by using technical specialists who do not always understand business and who do not effectively converse with business people. These factors exacerbate the changes caused by technology from the viewpoint of most business managers. Although technology changes are among the most important, they are among the most difficult to understand and control.

Change: New Opportunities to Win

The overall effect of a rapidly changing business climate is to increase competition and thus reduce profit. The high rates of change also produce new opportunities for market entry, which further increase competition. This may limit the performance of commerce overall, yielding smaller margins. However, change is not all bad; it provides new opportunities and it motivates the development of newer, better management practices.

An example of a new management practice that takes advantage of change is the use of increasing competition as a motivation. Confronting the entire workforce with the cost and quality of competing products directly motivates their support of quality and cost control programs. This approach is being used by many businesses today.

Like the other aspects of competition, change will be used as an asset by some businesses. These organizations will learn to control it better than others and will make it their primary competitive advantage. They will be rewarded for their efforts because change will be paradoxically a very reliable feature of this new age of business. These will be the businesses that survive, the businesses that win.

Business Processes: Work and Work Flow

Information and interpersonal relationships are important to business, but it is by process that work is accomplished. The business process is the basic unit of enterprise. It is the raw material from which the structure of business is built. For many reasons, process has received less attention from management science, from organizational science, and from business itself than the other views of business, such as organizational structure and information flow. Therefore, one of the critical problems uncovered when business process re-engineering is undertaken is that very little is actually known about process—a poor starting point indeed.

Process: The Essence of Business

Process can be viewed as the essence of business. Not only does most work get done through processes, but a great deal of what really differentiates companies from each other is inherent in their individual work processes. This is perfectly reasonable; the same raw materials and human capital are available to every business. Process is therefore one of the most important factors contributing to competitive advantage. However, despite the importance of process, it seems to have been largely ignored by management theorists and managers themselves.

What exactly is a business process? A process is most broadly defined as an activity carried out as a series of steps, which produces a specific result or a related group of specific results. Examples might be order processing or shipping. The term has also been defined as an overall aspect of the business operation, such as communications, which is performed within an organization. This definition was formalized by J. Kotter, in his *Organizational Dynamics: Diagnosis and Intervention* (Addison-Wesley: Reading, MA, 1976) and others, but this is not the definition most commonly used by business today. The steps of a business process need not be carefully defined, nor need they be consistent or carried out in particular sequence. They may be carried out by staff or by machines. The only critical factor is that processes are groups of actions that have a common purpose, and that purpose advances the business in some way. A process is generally larger than a task and is thus made up of tasks. It is normally considered to be smaller than an area of the business, such a operations, human resources, or shipping. The scope of each process is important only in that it should be a convenient unit to analyze, change, and manage.

Some of the best insight into business processes comes from a related topic: computer programs. Programs and processes have a common basis; both are a series of actions that are intended to serve business. They both have interfaces: places where they join with others and with staff. Computer programs, unlike business processes, have been subjected to much study, some of it highly organized. Although, it is not possible to assign all the characteristics of computer programs to business processes, there is enough analogy to be useful.

One lesson that programming can teach is that process is difficult to describe. For computers, data is easier to look at and has therefore received more attention. Computing has been much more successful at defining data than at describing the program itself. The relationship between business processes and organization is much the same. It is easier to describe the organization than the processes of any business. The result is that organization charts are common while process charts are rare. The results are also analogous: it is generally easier to find a bad data item than bad program code, and easier to look at organizational problems than process ones. It may seem cynical, but the most modern theories of quality assurance may be interpreted as just another way for managers to avoid involvement in process. Approaches such as total quality management advise managing people who do the work so that the workforce will fine-tune the work processes, without management's having to understand, analyze, and correct processes themselves.

Process: Target for Improvement

Processes are becoming the most attractive targets for improvements in business. They can be altered to improve:

1. Quality.
2. The business operation's efficiency and cost.
3. Customer service and response.
4. Competitive advantage.

Processes are improved for several reasons. First, they are the part of the business design that provides the most differentiation and potential for competitive advantage. Second, process improvement is the only opportunity to reduce costs significantly without reducing output or quality. Third, technology directly supports process, so that process improvement is the best way to take advantage of new technologies.

Because process is the framework in which other business elements work, other business improvements are best conducted within the context of a process renovation effort. Quality is a good example. A quality improvement effort that is attempted without changing process, will have limited scope. Both quality improvement and organizational change projects have been far more common than process alteration efforts. However, when process improvement has not been considered, the projects have generally been less effective than they could have been.

Not Engineered or Planned

Business processes have seldom been engineered or even planned. As already stated, they are usually developed informally over long periods. The only general exception is the industrial manufacturing process, which is almost always designed using industrial engineering techniques. Industrial engineering is a proven methodology, which produces well designed industrial processes. Unfortunately, it is not uncommon to find several of these processes, each optimized, existing in a highly inefficient overall business operation, along with inefficient administrative processes.

Business operations may lack design, but they have been nonetheless audited for many years. These audit reports and their associated working papers are among the best sources of information about the past and present state of operational efficiency and effectiveness in business. The findings of these reports are fairly consistent:

1. Few companies document production work flow processes.
2. Fewer still document administrative work flows.
3. Almost all business documentation is outdated.
4. Work flows undergo informal changes frequently.
5. Work of individuals depends on, and changes with, the work flow.

It is apparent from these audit results that the workers and line managers get their work done with whatever is provided. They change the work flow, tasks, and missions of their units as necessary to accomplish their work. That the changes are usually appropriate demonstrates their commitment, competence, and creativity.

The Boundaries of Work Processes

Since management is inclined to perceive the corporation in terms of its organizational structure, finding the boundaries of work processes can be difficult. Process interfaces may also seem to make the end of one blur into another. Certainly most business processes cross many organizational lines, from the bottom of the organization chart to the point where independent divisions are separated. Few processes, however, seem to cross division lines.

Because business processes are commonly in flux, their boundaries constantly move. This movement is greatest at the most detailed levels, of course. In many cases, the changes are discussed only at the lowest levels of management, some not even there. This flexibility is again the natural reaction of a concerned and creative workforce. However, the global impact of these local changes can hardly be assessed if no one is aware that they are being done. Senior management may, in some cases, see an effect but not be able to pinpoint the change in process that caused it. Process cost, quality, and efficiency are difficult to control, if changes take place in an uncoordinated manner. However, it is not desirable to stifle the good intentions of the workforce by stopping these changes. Management should, of course, be aware of them and provide control, but management should also encourage staff creativity.

Creeping Process Scope

Due in part to the extension of work processes across organizational lines, the scope of business processes, either overtly or surreptitiously, tends to expand. The movement is subtle; it can happen unnoticed. This phenomenon is normal in business, and it is also necessary. It is the principal balance to the tendency of business units to isolate themselves. This tendancy can, however, cause serious problems for change projects. The most common problem is that the spreading scope of the processes will place some of the process outside of the originally anticipated scope and control of the project. If this happens, the project's size will increase. These increases can be dangerous: they may result in a project that is both overly complex and too large to effectively manage.

Scope expansion can be controlled in several ways. The first is to carefully determine the boundaries of the business processes being changed. This may require a discussion on what activities properly belong to a process. If a consensus cannot be reached, the choice may ultimately require attention from higher levels of management. The key in selecting these boundaries is that the process definition must be used to confine the scope to a workable level. Of course, it is common to define and map, perhaps even analyze, more process than will actually be re-engineered. Somewhere, however, the limits of the process and the project must be formally set.

The other common method to control scope is based on computer systems development experience. Scope can be reviewed after the initial analysis and possibly altered, based on newly obtained knowledge. Both of these approaches are recommended in the use of Dynamic Business

Re-engineering, for which a special set of scope control procedures called "Amoeba Scope" has been developed (see Chap. 7).

Trying to Change Business

Business has been introspective for several generations. The works of such pioneers as Frederick Taylor and Frank and Lillian Gilbreth are now well over half a century old. The desire to improve business seems to have been kindled in the industrial revolution and accelerated by the technology explosion of the last century. This movement has culminated in a vast and diverse field of organizational change methodologies in the last decade. The direction of these new methods has generally been toward advancing change by influencing the workforce, instead of mandating changes in detailed work procedures. The theory was that the workforce, including management, could be expected to use its considerable adaptive capability to find and implement the best possible changes, if the proper organizational development methods were used. Furthermore, the implementation would be natural, smooth, effective, and robust if it happened from the bottom up rather than the top down. Most important, the workforce would automatically be highly committed to any changes that they had come up with themselves.

This seemingly progressive trend toward using organizational development methods as change agents has altered recently. Organizational change has become more prevalent, but it has also become less scientific. It has gone back to basics, and the reason is clearly that the urgency under which change has been conducted has not allowed the time required to learn new change methods.

Patterns of Change

Change in the past ten years has followed a particular pattern. The root of the pattern is the principal motivation for the changes: cost cutting. Generally, cost reductions have been due to increased competition and customer price sensitivity. The most common indication has been that the profits of the company have gone down in various amounts. Even in the cases of slight losses, the corporate response has been to cut costs, to protect the value of the corporation's stock, and to assure qualification for bonuses. This can be viewed either as a matter of business necessity, or as shortsighted and self-seeking, depending on the interests of the observer.

The first cost-cutting efforts seem to have been sensible, optimistic, trusting, and often completely useless. These first efforts generally took the form of requests from the executive level to voluntarily cut costs in forthcoming budgets. The response has been uniformly poor. The suggestions for cost cutting from middle management have been trivial, marginal, and almost always require other managers to give up more than the suggestors. Executive management has often been disappointed at the response to a request made in the best of faith.

The next efforts were often across-the-board mandated cuts in budget, typically ranging from 5 to 20 percent. The method used to make the cuts is more often than not some form of zero-based budgeting, resurrected from the 1970s for this special purpose. Unfortunately, the results have often further widened the gap between the executive and middle manager. The budgets submitted place the most fundamental, valuable programs in the least important budget increments, protecting the most vulnerable programs in the base level. This is a well-known and transparent device, almost insulting to senior management. Some budget cuts are made, but the effect is disruptive.

The final top-down effort is large-scale reductions in force. This is a desperate act, but most executives choose it when other avenues have been exhausted and the survival of the entire business is in doubt. Consultants are used to cut the workforce, and the effort is usually expanded to a full restructuring of the company. Corporate management allows itself to believe that the reduction will be a good thing. The staff who are outplaced will be the "dead wood" of the company and the company will emerge "lean and mean." These words are almost invariably used in this phase.

The results are usually so disappointing that few people will even discuss them objectively. Old-fashioned office politics, unions, and predetermined criteria, such as reductions in middle management, counteract any serious efforts to remove unproductive employees. Some of the best of the workforce are, not by coincidence, the most marketable, and many are lost when workforce cutting is first rumored.

Members of the corporate workforce and middle management exhibit their remaining defense mechanisms when confronted by what is to them a wholly arbitrary and unfair action. In many cases the corporate culture contributes to this reaction, and goes on to effectively defeat the attempts to restructure the company. When this has occurred, the business seems mysteriously to behave exactly as it did, despite the dramatic nature of the changes. Corporate culture is resilient; changing a number of people may have no effect on it at all.

The net effect of these attempts at cost reduction has been to make most of the companies that went through them less effective as competitors. One positive effect is that management no longer trusts simple answers to complex business questions. As a result, methods that more directly address the real problems, such as quality improvement and re-engineering, have been given increased attention in the past two years.

Resistance to Change

Resistance to change has grown out of negative experiences, unconscious reactions based on paradigm conflicts (see Chap. 3) and concerns that are at least understandable if not reasonable. Change of the magnitude required of business in today's new environment can be frightening, even to the bravest business person.

The competition among managers in many, if not most, corporations rivals anything intercorporate competition has to offer. This has induced the sad need to avoid any association with failure. Any manager who is acknowledged as having done anything wrong, even if the impression is false, will probably not survive long. This engenders an almost pathological fear of being associated with failure, or even being regarded as supporting something that ultimately fails. Since there is always some probability of less-than-complete success inherent in any change project, managers today must choose the ones that they support very carefully. However, managers also cannot afford to be thought of as being either unable to assume some risk or as the sort who reacts negatively to any new ideas; indeed, managers require the reputation of being visionary to succeed. Their resistance to change projects may therefore be subtle and even devious as a result.

Another commonly cited cause for resistance is the expectation of near-term external events. If a merger or takeover is anticipated, change may not be well timed. A closely related reason is the anticipation of management changes, especially high-level ones. A new president could be expected to halt a risky change and to form a bad opinion of those who initiated it out of ignorance. Also commonly cited is the possibility that changes in sales or in the marketplace may turn a bad trend around without having to make any changes.

Another common reason for rejecting change projects is cost. Because some projects can be very expensive, principally those for which new equipment or real property must be procured, many companies have difficulty allocating resources for them. Also, when urgency precludes advanced planning, the resources must be reallocated from other high-priority projects. Seldom is a provision made in any budget for projects that are not planned a year in advance. However, the rate of change in business today makes such long advanced notice the exceptional case. Resistance based on cost is most typical in those change projects that are not adequately supported by the highest level decision makers in the company. Without this support, the risk of failure increases significantly, and it may be better if the projects are not undertaken.

If the change project does not appear in the plan that accompanies the corporate budget, which is generally an annual plan (although there is a trend toward the use of three-year plans that are updated every year), then the company may not want to consider the project until a new plan is approved. Also, in some reported cases the preparation of new corporate strategic plans delayed change projects in order to avoid conflicts between the projects and the new plan. Again, the real reason for these deferrals is the lack of executive support. As stated, the change projects cannot succeed without this support, and a deferral in these cases is advisable.

Finally, a commonly cited reason for not implementing change is simple procrastination. In a number of cases, projects have been allowed to die quietly without ever being formally rejected or even openly resisted. They have been put off and referred to various committees and reported upon until it became obvious that the support required to initiate them in earnest is not there and never will be.

These patterns of change and resistance to change are the unstable foundation upon which any change effort in business today, including quality programs, will have to be constructed.

Where We Are Today in Business

Changes are necessary to meet the challenges and increased competition of the new global market. There is a growing awareness of where businesses must go to be more effective in this competitive environment. However, no sound schemes have been brought forward to help them get there. Also, there seem to be some significant impediments, regardless of the approach used.

Competitive Shock

The business world was taken somewhat by surprise by the sweeping changes of the last 15 years. It has always been difficult to distinguish between a new long-term trend and the short-term cycles. For example, the lack of real growth in established businesses was not seen as a real new trend until very recently, and these businesses remain unsure as to the extent of the global changes still taking place.

Clearly, many businesses have not known how to react to globalized competition. The opportunity and the challenge, even when recognized, failed to produce an immediate reaction in most companies. A few, however, have made actual moves toward taking advantage of the situation and have made efforts to expand their markets. The reaction of other companies has been limited to complaining about the lack of government protection and cooperation. Cooperation is desirable, but will take more time than the new business climate will allow.

In general, little concerted resource has been invested in finding out what is going on and doing something about it. Most executive business effort, in reaction to the immediate symptoms of the times, has been directed toward very short-term maneuvers, usually financial ones. Business seems to be waiting for some as yet undefined external events, or perhaps time is simply required to adjust attitudes.

This waiting period will end soon. The actions already under way will gather speed, as models of success are found and emulated. Today, however, even if the will to change exists, the methodology required to do so is not commonly known. The only factor that seems obvious today is that business must learn to change itself more effectively in the future than in the past.

Quality and Market Positioning

Next to cost cutting, quality programs and market repositioning efforts have been the most common approaches to change in recent years. Quality improvement has been attractive for several reasons: primarily it is seen as successful in some of the spectacular international industries and in some dramatic turnarounds of well established local ones. The consumer electronics producer, Emerson, is a good example. Furthermore, quality improvement, used to gain competitive advantage, was in the right place at the right time. It was one of the most important topics in front of management just at the time when the impact of global competition was recognized. Another reason for the pursuit of quality is that it is ineluctably a basic concern of business that has not been given proper attention and has not been used to its full advantage.

The problem that quality improvement presents is implementation. There is a limit to the degree that quality can be improved within existing, bureaucratic work process frameworks. Thus the most successful quality improvement efforts are in small and unstructured companies, in new companies, or in companies under the complete control of one person. There is also a limited potential capability to improve quality by motivating the workforce of a company. Significant gains can, however, only be made by re-engineering the underlying business processes.

Market repositioning is a direct, frontal response to increased competition. It would seem logical that any business considering change would at least review the position that the company and its products occupy in the market. The full range of corporate response to competition can then be assessed. Almost all businesses, from the smallest boutiques to the largest conglomerates, pay some attention to market positioning. However, there has recently been a trend toward special, intensive market positioning projects involving the highest-level executives in the largest established firms.

IBM is perhaps the best example. IBM began about five years ago to accept the loss of complete market dominance that it had enjoyed for a generation and to look for a different set of corporate roles in the future. IBM's response to change was diverse. The company bought minority interests in many small companies, reoriented its product lines, and established unprecedented cooperative efforts. This was a very open-minded series of actions for a market leader to take. The planning horizon was also unusually long; most of the results of these repositioning efforts will not be known during the tenure of the current executives. Furthermore, IBM seems to be continuing its repositioning efforts. Perhaps the analysis and improvement of the basic posture of the company with respect to the marketplace and to the rest of the business world will become a permanent concern of management.

As is the case for quality improvement, the limiting factor for market repositioning is implementation. Without improved methods, the repositioning effort ends with a desirable, but unattainable new position having been defined. Repositioning is a good beginning, but clearly not the whole answer.

Poor Decision Information

One of the most important aspects of business's current position is the lack of information on which to base decisions for change. This is not the same requirement as having the necessary data for daily business decisions. The information required for change projects is oriented toward how the company really works and what the marketplace will be in the future. The most difficult of this information to obtain seems to be the internal portions. For example, when examining the options for a given product, or for the enterprise, it is often found that the costs are not allocated as well as had been supposed and that the true costs of production cannot be easily determined at the detailed level.

Without having good information and figures related to the company's current operations, predicting the impact of change projects is impossible. The direct bottom line benefit of the project is only one of these impacts. All support requirements and the impact on staff, production, and operational activity must also be considered. Because these impacts are potentially negative and could offset the value of taking action, any decision support information must address these areas.

Unfortunately, the sources for change project decision data are not well developed in most companies; few have even defined their work processes. If the corporate business processes are not even defined, how can quantitative information related to them be gathered? Added to this problem is the lack of contact between the highest levels of the corporation and the lowest. The chief executive officer and board of directors typically will have to send representatives to obtain data from production processes—a time-consuming action. Because time is often a critical factor, they are generally forced to rely on the accounting function to provide the data they need. But the accounting data and production data provide very different views of the company—views that can be at odds with one another. For this reason, the use of either type of data without the other should be tempered.

In the absence of existing data, projects making changes in fundamental business practices will need to develop their own specialized sources of information. Unfortunately, this activity often takes much of the time allocated for these projects. To offset this problem, it would be helpful if companies began to gather process data on a routine basis. However, there is almost no movement at all in this direction at present.

The Need to Change Is Known

The overall impact of business's reaction to the new challenges of globalization has been limited compared with the urgency and scope of the reaction. The traumatic nature of cost cutting does not reflect the rather limited change that it makes in the company's ability to compete over the long term. The repositioning efforts to date have been limited by an inability to realize the changes that they identify.

It seems as though business recognizes that the time has come to make basic changes in the ways work is being done. Some attempts to change have been made, with partial and temporary results. The next step is to examine the paradigms of business to see how a static view of business operations can be replaced by a dynamic one, so that change can become an effective business tool.

3
The Paradigm Shift

Few discussions of business theory today are conducted without the mention of paradigms. The term has crept into businesses everywhere and can be heard from boardrooms to coffee break chats. It has become one of our top current buzzwords. But, like many such terms, "paradigm" has received more attention than understanding.

The dictionary defines *paradigm* as an example or pattern, especially an outstandingly clear or archetypical one. Joel Barker, a futurist, defines paradigm in his widely distributed videotape, *The Business Of Paradigms* (Charthouse Learning, Burnsville, MN, 1990), as a set of rules that establish boundaries and describe how to solve problems within these boundaries. Paradigms influence our perception; they help us organize and classify the way we look at the world. Taking this slightly further, a paradigm can be considered to be a model that helps us comprehend what we see and hear. It determines, to some extent, how we react to new information and can in extreme cases disable objective thinking regarding new information. One of the most important aspects of paradigms is that they operate on a subconscious level. In business, paradigms might be seen as sets of unquestioned, subconscious business assumptions. These assumptions, as we will see, certainly contribute to the paradigms of business people.

When the business world undergoes change, only those companies that react quickly will prosper. This ability to react requires considerable flexibility and an openness to new ideas and approaches. In creating this foundation the basic assumptions of business must be re-examined objectively and changed where appropriate. Therefore, to gain a competitive advantage today, the paradigms of the past must be exposed, reviewed, and changed to those of the future. This flexibility will not only be helpful in the near future; it will be necessary.

In his videotape, Barker discusses how paradigms tend to filter the way information is accepted and how they limit flexibility in considering new and different ideas. He contends that when we are presented with ideas that do not fall within the boundaries of our current paradigms, we have great difficulty in viewing them objectively.

From experience, it seems clear that these observations are correct. Paradigms seem to be a universal component of human thought; they will always be present and are not harmful in themselves. It is not the existence of these unconscious ideal models that is the problem; the problem is the limitations that business people allow them to impose.

This chapter looks at paradigms and their role in change. We will also consider the paradigm changes that must be made to succeed in re-engineering efforts. The concept of a new model of business, the change paradigm, is also introduced. It is an essential foundation element of the re-engineering method described in later chapters, and it represents a dramatic improvement in the way companies may manage themselves.

Paradigms: Resistance to Change

Change has always been resisted. There are numerous rational reasons for this: uncertainty, additional workload, risk of criticism, and interference with existing plans are a few. In addition to these causes, resistance can also come from irrational sources, ones that are difficult to identify because they have no obvious basis. Paradigms are often the cause of this unconscious resistance. If a proposed change clashes with a paradigm, the result will be a feeling of threat; a natural defense mechanism working on a subconscious level. The business person experiencing this feeling will then rationalize to defend against the threat, and a real problem will confront the change being proposed.

Paradigm filtering is also the underlying cause of many of our communication problems. Everyone has a different set of paradigms, so what is acceptable or obvious to one person may be rejected or misunderstood by another. The magnitude of this problem has been openly acknowledged for many years, but we have not been successful in correcting it. It may be that we have failed to recognize the role of paradigms in our attempts.

Paradigms set expectations. When reality fails to adhere to our rules in a given circumstance, we have difficulty understanding it. We may even reject a finding if it fails to fit within our rules. This often happens when we reject new ideas, without careful consideration: the door is closed to information and opportunity. In these circumstances, only a paradigm change will allow progress to be made. As the paradigm changes, so does perception and the ability to assess new information. In practice, advances are often made by people who are unencumbered with past

A senior management consultant began his career as an information services professional. After several successful computer projects, he was promoted high enough to meet regularly with the company's top management. In these meetings it was clear that senior management had no real understanding of what their information services managers were telling them. It was equally clear that the information services people didn't really understand the concerns and problems of the executive managers. Although a protracted dialogue was underway, little progress was being made and frustration was seriously impeding the business of the company. Fortunately, everyone involved respected the others and they all kept trying to communicate. However, no solution was found; it was a standing problem as long as he was with the company.

Later, while at a big six accounting firm, he crossed over to general management consulting and began to work directly for the presidents of companies. A big part of his job was to bridge the communication gap among the various groups in the companies. For the first time, he recognized the sources of problem. With few exceptions the roadblocks were perception and jargon. Everyone had his or her own view of the company and every group jealously held onto its own terminology—everyone was working from the foundation of his or her own paradigm. By bringing the groups to common ground in group sessions, a common understanding of each others' problems was achieved and they all understood how their decisions impacted one another. Through this understanding, the communication problem was temporarily overcome.

paradigms. Because they are unaware that something is impossible, they are free to find a way to do it.

Another paradigm-related problem that tends to resist change manifests itself as "not invented here." This occurs when the paradigm held by a group is very strongly exclusive of outside influences. To some extent it is an element of all group paradigms, a natural by-product of the group's cohesion. Again, the paradigm itself is not a negative factor, indeed it is the basis for group teamwork. The negative influence comes from the group's unwillingness to recognize the influence of the paradigm, and then set it aside or alter it enough to allow a gain to be made.

Assumptions

Assumptions and attitudes are among the most important components of paradigms. They are the subconscious beliefs that filter perception. Business paradigms in particular are influenced by business assumptions. The power of these assumptions to influence planning, decision making, and change vanishes when they are raised to the conscious level and examined. A few common examples of business assumptions are:

1. Business work is controlled from the top down.

2. Human resources activity should be separated from business operation management.

3. Good managers do not need to be experts in what they manage.

4. Jobs should be designed without reference to individuals.

5. Theory X: people must be forced to work, and Theory Y: people want to work and should be led, not pushed.

6. The organization chart is a real map of a business.

7. Managers ought to cooperate with their peers.

8. Each product or service should show a profit.

9. Corporate cultures exist, but are unimportant.

10. Unforeseen events will not affect business plans.

Each of these examples is discussed in the following paragraphs.

1. A basic assumption in business is that the work of the company is controlled from the top downward, with authority delegated to succeeding levels. It is further assumed that the delegation does not give subordinates the freedom to do the work as they think best; the work should be done as the senior person would want it done. This is not really delegation at all, but it is part of the paradigm and seldom questioned. Another indication of this assumption is that senior officers in business mandate actions and assume that the action will be carried out. This assumption seems to have been reinforced recently by more effective communications technology, since it has now become easier to exercise central control over geographically distant locations.

2. Another interesting assumption of modern business is that human resources issues are properly considered to be separate, corporate management issues. The details of this assumption include the definition of corporate standards for work and performance and the almost universal definition of corporate compensation classifications. While managers are assumed to be fully qualified to make work assignments, they are not assumed to be particularly expert in personnel issues. They are not assumed to be able to take disciplinary action, for example. Expert help is usually considered necessary for managers to conduct training and certainly to recruit.

3. Despite the assumption that managers need not, and indeed ought not, know anything about personnel work, the assumption is often made that a good manager can manage any part of the operation. Also, special technical qualifications are often not required for management. Many companies rotate managers from area to area, both to develop them for higher-level work and to bring new ideas to the areas of the business that might become stagnant. The assumption is that, as professional managers, they will be able to direct the technical knowledge of their subordinates without having any themselves.

4. Another personnel issue is the assumption that work, jobs, and positions in the organization should be designed around the work, not around the individual who is to do it. This is generally expressed as a policy, which is often violated. It is also assumed that, to ensure performance, jobs should be carefully designed as a detailed set of instructions. This assumption dates to the work of Frederick Taylor, considered the founder of management science, in 1900. Taylor used the example of loading pig iron to show that scientific job design can improve work efficiency, a concept accepted unquestioned by management today. However, it seems that Taylor's data, based on his study of Bethlehem Steel, was erroneous, possibly completely fictitious. His assumption may accordingly have no real basis.

5. The degree to which control over working-level staff should be exercised is another set of assumptions. The well-known Theory X and Theory Y, offered by Douglas McGregor, are not theories so much as they are assumptions of managers regarding the best way to manage. Theory X is the assumption that staff need to be closely managed and motivated by forceful and detailed direction. The Theory Y manager assumes that staff want to work and that they need to be led instead of driven. These two assumptions are the basis for others that distinguish two different basic approaches to management. The Theory X manager tends to make the assumption that all work should be specified in detail, which leads to the use of the Taylor industrial engineering approach, and to the use of top-down change management. The Theory Y manager assumes that the staff commitment and motivation are obtainable and important.

6. Another set of business assumptions is related to the organization chart. It is assumed, first, that there is an organization chart, although it is generally assumed that it will be at least slightly out-of-date. The organization chart is thought of as a map of the business, describing not only the formal reporting relationships, but also the functions being performed and the responsibilities of the company. Max Weber, who defined the ideal of a rational organization (*Max Weber: Essays on Sociology*, translated and edited by H. H. Gerth and C. Wright Mills, Oxford

University Press; New York, 1958), included among its characteristics a "well ordered system of rules and procedures that regulate the conduct of work." Most business people today continue to consider this definition to be true.

7. It is assumed that there will be peer-to-peer cooperation. This assumption is believed to be theoretically true because it is clearly in the best interest of the business. It is, however, not supported by many corporate policies, and there is at least as much motivation for withholding cooperation as for rendering it.

8. Some very basic approaches to profit are also influenced by assumptions. There is an assumption that each product or service of a business should pay its own way. This assumes that product and service costs will be carefully determined and monitored. In some industries, cost accounting is indeed considered very important, but in others it is not. Some companies have no accurate figures for their product costs, preferring to manage by overall cash flows. However, even these companies rarely state that they do not consider the costs of products to be unimportant.

9. Another question of importance is the assumption that corporate culture, which is widely acknowledged to exist, is not very important. The common assumption is that every company has a culture, which is composed of its own ways of doing things, its own assumptions, its history, and even its own language. However, it is also assumed that these things do not influence the way the company does business, and they could all be changed if there were some reason to do it.

10. Perhaps the single most important assumption, in terms of corporate attitudes and plans, is that there will not be any unforeseen external change pressures over the corporate planning horizon. Business certainly tries to plan for predictable changes, but seldom makes real contingency plans for externally forced changes that cannot be predicted.

Are these and the many other assumptions made by business today valid? If not, are they serious problems or impediments to success? It is not possible to generalize. In most cases, assumptions hold, or they would be discarded. However, in most planning efforts the identification and confirmation of assumptions are considered worthwhile. The assumption that is not tested is the one that may cause the entire plan to fail.

In re-engineering business processes, testing assumptions is especially important. Re-engineering analysis looks at the current process, the work that must be done and the parameters that constrain the process. It then attempts to bring about a new process—to define a new start to replace the old, patched process. To gain the advantage of a new start, the basic assumptions of the business must be discovered, questioned, validated, and tested to assure that the new process design does not carry over unnecessary effort. Experience has shown that re-engineering has its greatest chance for success when nothing at all is assumed.

Attitudes and Paradigms

Personal attitudes are closely related to paradigms. Our attitudes, not our paradigms, determine our willingness to change. Attitudes are a combination of our individual personalities and our experiences. As such, just as with paradigms, they evolve. But unlike paradigms, which are sets of rules, attitudes are much more nebulous. Therefore, they are much more difficult to change. As our paradigms act as filters, our attitudes color and shade how we view things. Literally, our attitudes govern how we will use our paradigms. For this reason, we need to consider both the rules we use to understand our world and the attitudes we have developed throughout our lives when we look at change.

Paradigm Changes and Paradigm Shifts

A *paradigm change* or *shift* is essentially a significant change in the rules, assumptions, and attitudes related to an established way of doing something. The term has also been applied to a fundamental change in a technology, to emphasize the impact of the technology's new capabilities. A paradigm shift has the effect of a new beginning. Past success does not guarantee success in the future. In fact, past accomplishment may be detrimental if it causes the rejection of new opportunities and resistance to change. It is possible to be so set in the ways that have worked in the past that it is impossible to recognize a changed situation, consider a better way, or take advantage of a new opportunity. However, when a significant change in business takes place, the old paradigms must change and allow the consideration of new actions. What was impossible yesterday may well become commonplace. If a company fails to take advantage of these changes, its competitive position will be decreased as the competition moves to exploit them.

So the future cannot be viewed through current paradigms. It must be recognized that the ideas and techniques that succeeded in the past may not be the ones that will take a business into the future.

Foundations for Paradigm Shifts

Businesses go through a common evolutionary cycle, as described in Chap. 2. One result of this evolution is the growth of the supporting infrastructure of the company. This infrastructure includes the elements of the business that are shared by all of the organization. These infrastructural elements support, define, and regulate the work. Corporate management, the human resources function, information systems, telephones, the physical plant, and accounting can all be considered infrastructural. Corporate plans, the business architecture, and the design of business processes (even when not documented) are also part of the infrastructure. As businesses become more complex, individual managers and staff add to the real work processes by making ad-hoc changes in response to problems and changing requirements. This creates an informal infrastructure, which over time diverges from the written rules, organization structures, position descriptions, policies, and procedures. As a result, any documentation that may exist is normally incomplete, inaccurate, and out-of-date.

This reluctance to spend the time necessary to formalize and update corporate documentation has been pervasive. Few companies have formal business rules. Many are implicit or just known to those who do the work. Activity, policy, and rules are thus subject to individual interpretation. Also, as one moves from department to department, the informal rules change significantly. The result is that it is not possible to obtain consistency, and the outcome of any activity can be expected to vary.

In one company that the authors helped to move to the change paradigm, the company's rules were contained in a set of manila file folders. They consisted of memos dating back to the mid-1960s. When asked if they were used in training, the response was no. When asked if anyone looked at them, again the answer was no. So we analyzed the memos. Our findings were not surprising. There were contradictory memos and rules about rules. Because our approach associates business rules with the operation they direct, we needed to sort this out. What we found was that there were two sets of rules: the ones in the manila folders and the informal rules that ran the operation. We documented the informal rules and combined them with a list refined from the ones in manila envelopes. We then examined each rule and, working with mid-level and line management, helped reach a consensus regarding the rules that should be formalized. We then applied the applicable rules to each person's job and worked with the staff to determine how their tasks would change. The result was that the quality and consistency of their work dramatically improved. Where management could not count on the same response to the same question or action from any two staff members, they now had consistent outcomes.

Because corporate paradigms related to company functions are influenced by rules, policies, and procedures, the lack of formality has been a key factor in the ever-present communication problem. Corporate managers believe they have a common terminology base, but often they really do not. Departmental variations of seemingly standard concepts, rules, and terms are common. Each department, and often each manager within a department, views and interprets rules and policies differently. Their business operation paradigms are thus different. These differences cause a lack of universal focus in companies, which adversely impacts consistency, quality, and efficiency.

This lack of formalization has seldom been viewed as an important problem. Any company that is currently in business has obviously been able to perform all the work that needed to be done. Continuing to do the company's work does not seem to require formal work process design. The problem comes when changes are required. Most companies today find that changes *are* required—indeed, urgently required. But, although change is required, it seems impossible to achieve; there is no foundation for it. For example, technological advances are changing what is possible, but the lack of a baseline makes implementing technology pure guesswork. Companies simply must be able to recognize opportunity, design a solution, determine its impact on the operation and then execute significant changes within months—not years. The question has become how to survive—how to compete effectively.

Another key component in the now recognized need to change is the inertia that has grown over past years. Even companies that have historically resisted change have experienced some movement. These changes are generally made slowly and quietly; few people are even aware of it. Slow movement provides comfort and a type of stability to the organization, while allowing the company to react to new situations. Unfortunately, this practice is also the cause of companies failing to document their procedures, and it is the reason that management seldom has a detailed understanding of how work is really performed. This slow, inertial movement makes the foundation for re-engineering a slowly moving target, all the more difficult to deal with.

Why Change?

Where competition is low and business is good, a company will reject significant change and continue to do business as it always has. It will not evolve. It will not reinvest. There is really no reason to "rock the boat." If it isn't broken, don't fix it. If management wants to increase profit, it will simply raise the price of the product. If inefficiency and waste creep into the processes, they will be handled by increasing staff and passing the cost back to the customer. Certainly there are limits to this behavior, but the companies that have no real competition seem to reach these limits and gradually push them higher.

When competition arrives, the picture changes. There is a paradigm shift. Companies that make the transition to the new paradigm succeed. Those that resist may not.

Today there is just such a marketplace paradigm shift. The business operation must be changed as market pressures force companies to respond. The rules of the past are being rewritten and survivors must recognize and accept the new rules.

The Ripple Effect of Paradigm Changes

Observation has also shown that paradigm shifts in one area cause ripples to occur in our other paradigms. Change to the rules of one area inevitably cause changes to other related rules. This effect continues as the impacted rules, in turn, impact other rules.

The Limitations Imposed by the Old Paradigm

Each company has its own set of operational, social, and technical paradigms. These ways of interacting and doing business define how the business runs today. And, while these factors provide a firm base for the

---

For example, in the hypercompetitive global market, a key differentiator is quality. Business will accordingly reassess what quality means and how it can be measured. The old paradigm sees quality as an internal issue, to be improved by raising standard and measuring success by inspection. The new marketplace, however, will not allow that narrow view: quality improvement must now have an external component. Externally, the customer's opinion of quality is what is important. Customers must have confidence in the product. To obtain this type of quality, business shifts its paradigm of quality to include a strong emphasis on the customer's view of quality, as well as the competition's view. This definition of quality will start with management and, in turn, cascade through the production operation and the customer services operation—through the whole company in time.

company, they can also stifle progress. To help avoid potential problems, the following discussion deals with the negative side of these factors; the ways in which they can hamper a re-engineering effort.

One of the biggest impediments to moving forward is corporate culture. Culture (discussed in more detail later in this chapter) is a limiting factor related closely to paradigms. An example of a cultural limitation is the relationship between the company and the workforce. In today's business climate of cutbacks and "right sizing," both managers and staff have lost faith in their companies. This situation is one of the driving factors behind low-profit management and reluctance to present new ideas. Many if not most, managers today are concerned about the stability of their jobs. They believe that they are constantly being pitted against other managers and feel that the risk of rejection is to be avoided at all costs. As a result, valuable ideas on improvement are withheld.

The second most significant impediment to a streamlined, efficient operation is technology. This includes production equipment as well as computing and communications technologies. The difficulties presented by technology are all well known. Applying technology to business problems is difficult. Using technology to its greatest advantage seems an impossible goal. The real benefits of technology are impossible to determine. And, to complicate the other problems, the costs, time, and risks involved prohibit making widespread changes to corporate technology investments.

Together, the culture and the technology of a company provide a profile of its environment. Any analysis must consider this fact. Any proposed change must also recognize this profile and blend it into the design and presentation of alternatives. Also, a proposed change must recognize the interface requirements and the limitations that the current technologies impose.

The third significant impediment to change is instability. Company management and often ownership change too rapidly to provide long-term stability. The changing policies that result and the fear of taking actions that may be criticized or reversed by new management are almost constant factors in business. New initiatives are particularly vulnerable to this problem, losing both time and momentum. This factor also severely restricts any attempt to engage in long-term efforts, such as re-engineering or moving to a new business paradigm.

The final issues that impede change are time and commitment. The problem is that few managers will willingly commit the time or resources to become adequately involved in change projects. Because it is assumed that there is fat in the operation, senior managers mandate the involvement of mid- and lower-level managers in remedial activities. However, mid-level managers may only make a show of support. Because only the best people should be involved in corporate change projects, the services of valuable staff are requested, but line managers blanch at the prospect of losing a critical person. As a result, the project staff becomes over-worked as they must work with the project team while continuing their original jobs. The alternative seems to be assigning the dregs of the company. Not a pleasant thought.

Casting Off Old Ideas, Assumptions, and Rules

It is important to cast off the aspects of current paradigms that stand in the way of progress. This does not mean doing away with rules; it is not casting off management in favor of anarchy. Business must have formal policies, rules, and procedures. It is the impediments to change that must be eliminated.

New attitudes must replace old ones. It is, of course, more difficult to change an attitude than a rule. To accomplish this, it is necessary to understand the reason for the attitude and work to change the underlying factors. For example, some managers fear change because they see it as a threat. They fear disapproval and failure. By removing the stigma associated with taking risks and potentially failing, companies can open their people to new ideas.

Another obstacle that must be overcome is the feeling of ownership some managers develop for their areas of responsibility. They may perceive a review of their operations as a personal threat. These managers are normally afraid that any proposed improvement will be regarded as a negative reflection on them.

By removing the assumptions and attitudes that are the causes of resistance to change, it is possible to take a new look at the business. The general remedy is to overcome the distrust of the motives of senior management. Clearly, this will not happen quickly. Trust requires evidence. Evidence requires consistent commitment to a goal and consistently applied policies.

To make a habit of change, a company must demonstrate commitment through action. This commitment has unfortunately been one of the casualties of business in recent years. Quality and cost containment are the results of operational improvement and require a long-term commitment. The creation of an environment that can change rapidly and respond to new opportunities also requires a long-term commitment. These objectives must be the foundation for a long-range strategies. And, although nothing can transcend an acquisition or merger, this strategy must transcend management changes.

An example of old assumptions that should be cast off is the current methods for the evaluation of proposed business changes. Investments are usually analyzed using financial standards, return on investment (ROI) being the most common. ROI attempts to analyze the full cost of a new investment, but seldom includes the costs of organizational changes and business process changes, because these costs are difficult to predict. Changes that do not result in an immediately definable ROI are normally addressed on a case-by-case basis. These change efforts are the most difficult to decide on and to implement. The assumption is made that these changes cannot be analyzed. It is also assumed that these projects will happen infrequently. If the investment paradigm is consciously examined, it can be changed to accept the need to analyze and implement such changes on a routine basis. The company can establish methods to evaluate noninvestment change proposals by modeling or by other appropriate means. Corporate planning staff can change their procedures to accommodate all forms of change proposals, not just the ones for which the analysis tools are convenient and comfortable. Many new ways to quantify goals, expenses, and performance can be introduced after the assumption that they do not exist has been discarded.

Re-engineering: The Movement to a New Paradigm?

Business process re-engineering, hailed by some as a new paradigm, is not a new paradigm in itself. It does, however, *require* a new paradigm to be effective. This new paradigm is a willingness to question everything constantly. Clearly, trying to re-engineer without challenging our basic assumptions about business will not bring the desired results, a failure that many early attempts have demonstrated.

While business re-engineering is fairly recent, attempts to streamline business operations and improve the efficiency of business in general have been common for a long time. For example, the efficiency experts who did time and motion studies during the first half of the twentieth century were occasionally laughed at, but they were often effective. Their work led to the formal study and practice of industrial engineering. The move to computer support in the 1960s was also intended to improve productivity and efficiency. It was believed that both would reduce costs and would result in improved customer service. Both industrial engineering and computing succeeded, up to a point, and continue to be useful. Specific operations were often improved and costs were reduced, but the problems associated with complexity, organizational growth, and bureaucracy continued.

A common view of paradigm change is that the paradigms of business have shifted due to the new popularity of total quality management, the trend toward internationalization, the increased use of technology, and the step-up in competition in nearly every economic sector. These factors indeed motivate a re-examination of the basics of business, but in most cases the paradigm has not shifted. Common misconceptions exist regarding total quality management (TQM) being a new paradigm, which it is not. The basic business assumptions are generally not changed by implementing TQM; in fact they are usually strengthened. TQM is certainly new and it is very effective, but it is a performance-enhancement method, not a careful re-examination of basic assumptions. Although re-engineering and TQM are steps in the right direction, there is along way to go.

Re-engineering: Characteristics That Support Success

Why now can the business world claim to be better at promoting change than in the past? Why can business now succeed? The answers lie in the advancement of knowledge, in the understanding that time, trial and error, and advances in technology have brought, and in the recognition that short-term solutions are not the answer. Looking at past efforts to bring about change in business using traditional methods (such as reorganization), the following problems can be identified:

1. Companies did not have either the formal documentation of current practices or the infrastructure to use as a base for change analysis and improvement design.

2. There was insufficient real management support for change projects.

3. Long-range business strategies were unrealistic and had insufficient influence on both new planning and routine operations.

4. There was insufficient understanding of the capabilities, limitations, and costs of technology.

5. Information services and technology strategies, where they existed at all, were not cohesive and comprehensive, and were largely ignored after being written.

6. There was no understanding of the relationship between the business operation and its computer system support.

7. It was common knowledge that, if the effort succeeded, people would lose their jobs; cooperation was often minimal (efforts were resisted by staff and managers alike).

8. Change efforts were very narrowly focused and were limited by organizational boundaries.

9. The concept of business processes was not understood.

10. It was believed that a perfect solution could be designed and then implemented as a single large project, the results of which could be expected to remain effective for a long time afterward.

Then came re-engineering. However, many of the re-engineering efforts of the last few years show that the problems of the past continue. The similarity between the problems experienced before re-engineering and the problems of early re-engineering projects indicates that the new method was being attempted without a change in paradigm. The authors have observed that past projects that have been called re-engineering:

1. Approached re-engineering as an operational review effort.

2. Tried to create a perfect operation in one move.

3. Defined the scope too narrowly.

4. Re-engineered the organization instead of its business processes.

5. Failed to obtain the right information.

6. Failed to understand the ripple effect of change.

7. Did not adequately consider culture.

8. Failed to consider technology limitations.

9. Did not commit the required resources.

10. Failed to model alternative solutions.

But as business steps up to increased competition, the need to succeed at re-engineering is greater than at any time in the past. To succeed, however, the paradigms regarding re-engineering projects must themselves be changed. The new business re-engineering paradigm is based on the following principles:

1. Quality can only be achieved through a process of continuous improvement; it is futile to attempt to reach perfection in a single step.

2. Change itself should be regarded as a never-ending process; once started, a company should never stop evolving.

3. Proposed changes can only be evaluated as the difference between

the current operation and a new operational design; they should be evaluated by modeling.

4. Change efforts must be based on a detailed understanding of the company's processes.

5. Quality initiatives can be implemented only by building them into the fabric of each process.

6. Efficiency and cost reduction are truly gained through the reduction of waste.

7. The modeling technique used must support dynamic modeling; models must be able to continuously change in a controlled manner.

Furthermore, to succeed, a re-engineering project must be based on:

1. A firm long-range commitment from senior management.

2. An understanding of the process and work flow of the company, along with the identification of the interrelationships among department.

3. Information related to the business processes answering the six basic interrogatives: who, what, when, where, how, and why?

4. An understanding of the corporate strategy, its goals, and its problems, for both the corporation and for each department.

5. An understanding of the responsibilities of each department.

6. A definition of operational and production problems.

7. The use of fluid models for the operation.

8. An understanding of change and how it can be used as an ally.

9. An understanding of the current technology and its status, including (a) production equipment, (b) communications equipment and networks, (c) computer equipment, and (d) computer software and data files.

10. An understanding of the corporate culture.

11. An understanding of the ripple effect of change and the ability to predict the impact of all changes.

The Same Old Thing

The concepts of product and work quality, as well as of operational efficiency and effectiveness, are not a new paradigm. They are objectives that drive many companies. The move to a new paradigm is related rather to the manner in which companies attempt to achieve these objectives.

So today most re-engineering is "the same old thing in new clothing," because these efforts are approached within the framework of the old paradigm. But it should be emphasized that most re-engineering efforts produce satisfactory results. The problem again is that too many fail to move the company into an environment that prepares the operations for long-term gain. The approach discussed in the following chapters offers exactly that advantage.

The objectives of the re-engineering effort must also be refocused. The needs are different now that companies must gear for constant change to remain competitive or to respond to the ever increasing burden of government regulation. Therefore, in addition to immediate gain, which each re-engineering effort must provide, these projects must also provide a sound basis for future operational simulation modeling and impact analysis.

In approaching these efforts, everything must be questioned and justified. There can be no sacred ground and there can be no accidental omissions. The molds of the past must be broken as a starting point. By doing this consistently, the natural reluctance and fear of criticism can be negated. Where this is not done, the company runs a very real risk of doing exactly what the early computer systems did—"paving the cowpath" instead of looking for a better route.

The Second Paradigm Shift

The first paradigm shift, while largely undefined, entered the scene with quite a bit of fanfare. It was the vanguard of the current move to streamline businesses. It seems that the paradigm shift was based on a recognition that it is necessary to improve quality and operational efficiency. Unfortunately, the attempts to support this shift have been based on the old approaches, using the same techniques that have failed to produce acceptable results in the past.

We believe that further changes in this first paradigm shift are now occurring. While the new business paradigm that will result will have many of the same goals as the earlier one, it will also recognize relationships and requirements that the first paradigm shift did not consider. The most important difference is understanding the nature of change.

Although most of the important aspects of change are well known, the ability to use them effectively is not. Chief among these is the ability to use change continually. Also, the ability to change rapidly is needed to win competitive advantage. The second paradigm shift will thus result in the development of approaches to re-engineer businesses based on the concept of continual, managed change. We call this newest set of business concepts the Change Paradigm.

The Change Paradigm

A high-priced consultant and many classical philosophers have said that
the only constant is change. And with the improvements in travel and
distribution, the international nature of business, and the advances in
technology, change is coming faster all the time. As marketplace, government, and technology continue to apply pressure, business activities,
methods, and operations will need to be reviewed and modified constantly.

But, unlike the past, where significant changes were few, the modifications of the future are likely to have a high impact on the operation. This
intensification is the result of greater complexity in companies today and
the ripple effect of change. To deal with these changes, companies must
create a new infrastructure. They must become very flexible and capable
of quickly evaluating opportunities and alternative approaches, and then
reacting within a very few months. This ability to react fast, with high-
quality, cost-effective products and operations, will provide a significant
competitive advantage.

The *Change Paradigm* is an approach to business in which the operation is oriented toward a continuing change. It is based on the belief that
quality and efficiency can improve only through a constant evolution. In
this environment workers and managers are expected to question everything and seek new and improved ways of doing things. All rules and
work are constantly scrutinized. New ideas are encouraged. And the
evaluations must be carried through to the implementation of the
appropriate modifications.

This paradigm is also applied to the fundamental concepts and products of the business. Management must continuously evaluate the
reasons for continuing to compete in each market and within each line of
business, and it must be open to investigating opportunity. This evaluation is carried to the detailed levels of the operation where each policy,
rule, procedure, and task can be reviewed and either justified or deleted.

In this way process redundancy and unnecessary rule overhead are
eliminated. Everything that is done will eventually be related to a specific
purpose and every purpose related to a business objective. Through this
evaluation, management and staff will gain a firm understanding of the
value of each task that is done. Everyone will also gain a clear insight into
the real operation of the company.

Operating within this paradigm represents a commitment on the part
of the company to create consistency of purpose. It also represents the
acceptance of the belief that quality improvements are evolutionary in
nature. Operation in this paradigm also requires a commitment in concept from senior management. Because it represents a new approach to
business operation, it will need strong executive backing. Since everyone

in the company will ultimately need to accept the concepts of the paradigm, senior management must constantly endorse this direction. Meetings should begin with the question of what is changing and why. Some companies will need to go so far as to establish a change officer and to support this position with both budget and staff.

A large part of operating in the Change Paradigm involves an understanding of the need to formally model and control change. These models (discussed in the following chapters) are the foundation for the simulation modeling, impact analysis, and cost estimation that are critical to evaluating new ideas and designing the most effective implementation scenario.

These models begin with the current operation—"current" meaning the present operation at any point in time. Change is then defined as a modification of the existing rules, policies, procedures, processes, and the like. As each change is implemented, the resulting operation becomes the current one, and the next subsequent change is applied to that version. In this way the operation evolves.

To operate within this paradigm, a company first defines its infrastructure. Models of the current operation are created and processes clearly defined. These models, along with the supporting data on who, what, when, where, how, and why, allow management to simulate a new set of procedures that are predicted to improve the quality of a process. They allow management to understand how the change will really work and who will be needed to make it a success. As each low-level action is associated with higher-level processes, it is possible to determine the full impact of a proposed change.

The Change Paradigm is a conceptual environment. The approaches, techniques, and tools discussed in the following chapters provide the ability to create the necessary corporate infrastructure. Once a company begins operation within this paradigm, the process of reengineering never ceases. It becomes constant but incremental, as the company evolves toward better quality and efficiency. This represents a new business operation life cycle.

The New Business Operation Life Cycle

As a result of operation within the change paradigm, a new business operation life cycle is emerging. This life cycle is very different from those of the past in that it combines business operation, production, information services, and communications into one integrated whole. While this relationship has not formerly been stressed, the need to consider the interaction of these components has become apparent. The second major difference of this life cycle is that, once started, it will not become obsolete and end with the replacement of the "operation."

Once started, the life cycle of any effort becomes dynamic. This is the difference. In the past, activities had clear life cycle stages: conception, gestation, birth, growth, maturity, and senescence. Eventually, the cycle of the typical business operation brought the activity to the point at which it needed to be replaced, and then the cycle began again. Business re-engineering is often thought of as a new step in the old-fashioned life cycle. The re-engineering step is supposed to bring the operation to a new birth from any other point in the cycle. After the re-engineering project is complete, it is assumed that the life cycle will be played out as before, until the next re-engineering project is initiated.

For those operating in the change paradigm, re-engineering represents a constant evolution of the operation toward perfection. This continual process has a definite beginning, but, because of its constant use as an enabler, the process has no end. Activity levels do, however, decrease significantly following the initial re-engineering project.

Starting with a Clean Slate

Joel Barker, in *The Business of Paradigms*, states that when a paradigm shifts everyone starts over again. Certainly, when significant shifts occur, those who can recognize the opportunity and take advantage of the shift outperform those who do not. History abounds with examples. The stirrup gave mounted riders a tremendous advantage in battle. This simple invention allowed cavalry to become effective, and it changed warfare.

In business today the same relationship of advantage and paradigm shift holds true. Railroads—not understanding that they were in the transportation, not the train, business—failed to recognize the true potential of aircraft and have paid a high price. A modern-day example is the oil industry. If they fail to recognize that they are really in the energy business, they may miss the opportunities of profiting from alternative energy sources.

These examples are important only because they represent failures to see beyond the current paradigms. They show only too well what can happen when people fail to remain open to new ideas or inventions. And they show that, when a significant technology or idea changes, everyone involved with that paradigm starts over. The Chinese did not benefit from being the inventors of gunpowder. The railroad industry failed to develop air travel and lost much of its market. Clearly, those that miss the paradigm shift may and often do lose.

But becoming involved early in a paradigm shift is very risky. The paradigm may shift differently from how it is anticipated to. Future paradigm shifts can often be seen in new technology or opportunities, but the outcome of the shift is often dependent on the creative application of the technology. This creative application is where the risk and market position dominance occurs.

The opportunities these shifts present are limitless. They offer a clean slate for the creative application of new techniques, materials, and processes. And because some of the largest companies involved in the current paradigm may be hampered by closed-minded attitudes and rigid paradigms, their current dominance of a market may be taken away by a more innovative and much smaller company.

What Is Quality?

As shown in Fig. 3.1, "quality" means something different to almost everyone. To encompass these varying views, quality must be given a composite definition. Furthermore, this definition is useful only when there is consensus. Today, many companies begin re-engineering and

Figure 3.1. The differing views of quality.

quality programs without a clear understanding of what quality means within their own particular industry or activity. If this happens, the effort may be fatally flawed.

A team of people from a company, its customers, and its competition is the best group to define quality. In practice, no one of the three components of this group can do the job alone. Furthermore, this working definition of quality should contain specific standards at two levels. First, the managers of the company should set the framework for quality based on their understanding of their marketplace, their customers, and their competition. Next, quality should be defined for each line of business. Again, the definition will be drawn together from the customer, the workforce, and management, as well as from what the competition is doing.

Like most actions within the Change Paradigm, the definition of quality will be a moving target; it will change constantly as the views comprising it change. This fact has far-reaching implications as operations, production, strategy, and service requirements are tied directly to this definition, and it will need to change in response to definition adjustments. Accordingly, at least a quarterly review of this definition is needed to stay responsive to marketplace demands.

Continual Quality Improvement

Continual quality improvement (CQI) is a process by which quality improves constantly. Following W. Edwards Deming's principles, quality can be improved only as part of an ongoing improvement of the work processes of a company. This requires an understanding of the company's processes and the implementation of a total quality initiative (TQI): a commitment to CQI at the most senior level of the company, the marshalling of the company's resources, the education necessary to modify the corporate culture, and the investiture of quality as the primary goal of the company.

Although there are many approaches and a great deal has been written about this topic, few organizations have established a true quality improvement program. Among the chief reasons is that most businesses lack the infrastructures required to implement a TQI program.

The starting point of any quality initiative is to understand the business operation as it truly exists today. This understanding cannot stop at a conceptual level; all managers know basically how the company should work and how the company's operations fit together. But few managers know the details of how the operations *really* work—what each department does, what each person does, and how it all fits together.

We have found that to move into a TQI program, managers and staff must understand the operation at a very detailed level. They need all available information on the interrogatories of who, what, when, where, how, and why to address issues and evaluate proposed changes. In addition to the operation, this understanding must also include the support, production, and communication systems. Only when viewed in a composite whole can the operation be understood, and change controlled within it.

An understanding of processes is critical. Processes are almost always fragmented, typically divided among several departments. We also know that, because of the relationships between processes, a change to any part of a process will affect a great many other processes. Without this understanding, what can safely be changed? Projects that are not based on a real understanding of the processes and their interrelationships can affect only part of what they must. They cannot address the problems and opportunities for improvement in all parts of a process, they cannot predict the result of a change, and they cannot anticipate the ripple effect.

Today, most quality initiatives fail to support key factors in this concept. Most companies and consulting firms approach quality initiative as a single-step effort which, once implemented, will provide quality for the next several years. Also, the majority of efforts have been departmentally oriented, and not process oriented. Finally, most efforts continue to rely on quality measurement as the key to quality.

Operating within the Change Paradigm removes these problems and allows a company to implement a total quality initiative program. When an operation is set to change continuously, the evaluation of the processes and the products of these processes can be used to feed a continual processes improvement program. This program provides a vehicle for the evaluation of all quality assessment findings, ideas for improvement, problem resolutions, changes in the marketplace and new regulations. Given simulation modeling, all the ramifications of these motives for change can be assessed and the best alternatives selected.

When moving to an operation based on a continual quality improvement, a series of smaller focused actions is used to improve quality. The processes change over time, not as the result of a single action. CQI is thus a commitment to changing in a controlled manner, forward from the point it is adopted.

The Importance of Corporate Culture

Every company has a culture. Cultures are based on the most deeply seated paradigms of the company, and they provide a background that colors all the actions of the workforce. Culture exists below the level of written policy and procedure, arising from elemental attitudes. Most corporate cultures pervade the entire enterprise, with some variance among divisions if they are widely separated and independent.

Culture has its strongest impact in two areas: interpersonal relationships and change. In these areas it is so influential that new approaches to each must account for existing culture, or they may easily fail to be effective. This is one of the chief complications in applying new management technologies. Any management methods that do not specifically consider culture tend to work well in some cultures and fail in others.

Categories of Culture

There are no generally accepted categories for corporate culture. It is more appropriate to discuss culture as having certain characteristics, with each organization's culture being a mixture of degrees, although some characteristics do not mix well with others. A few of the terms used to describe culture are as follows:

Open: Organizations in which mobility is high, levels informal, and Matrix management or teams are common.

Formal: More or less the opposite of open; levels are rigidly observed, there is little interaction across them; rules are inclined to be written.

Progressive: An organization that likes to move forward, and try new things.

Political: A clannish organization; either open or formal, but with management that allows, possibly encourages, the formation of informal alliances that make business decisions outside the formal processes, but not by consensus.

Entrepreneurial: Like a political organization, but without the alliances; every person for himself or herself; initiative is encouraged; initiative can be either constructive or destructive; generally more likely to be open than formal.

Family: Like a political culture, but the alliances are more or less permanent; they need not be real families, but they are in some cases.

Other cultural characteristics include the relationship between the workforce and management, such as the labor union, which imparts a certain set of cultural parameters, and national cultural factors.

Culture and Change Projects

Cultural impact on change projects can be illustrated using some of these characteristics. One example is the impact of a simple corporate policy. If a policy is issued in a formal culture, it is assumed that it will be followed, and it usually will be. If the president of a bank, many of which are formal cultures, issues a directive that restricts the procurement of personal computers to a specific brand name, it will be effective. In any of the more open culture types, the president probably would not issue such an order, but if it were issued, it would not necessarily be followed. Requests might be made, and granted, for exemptions to policy in those cultures, or in some companies the exemption would be assumed. These approaches differ dramatically, and they will affect re-engineering projects profoundly.

The other aspect of culture that is of concern to change projects, and particularly to the process of positioning, is that changing corporate culture may become one of the goals of the project. It is unusual to find that the basic culture of a business or institution is inappropriate to its business ambitions, but it can certainly happen. When it does, it can be a serious impediment, especially if the cultural problem is not identified.

In cases where many changes are tried, but the same conditions seem to prevail, there is often a cultural problem. Cultural problems are very resistant to personnel and organizational changes. If the problem is identified, the corporate or institutional culture (institutions seem to have greater problems with this than business does) can be purposely changed, but only with difficulty and expert help. Later chapters offer some hints on dealing with culture in re-engineering projects, but very experienced and specialized assistance is recommended when serious culture problems are being solved.

Quality, culture, organization, infrastructure, and even business processes contain the unconscious but highly influential elements of business paradigms. Fundamental to progress toward a more competitive position is the ability to examine, understand, and change these most basic foundations of business thought.

4
The Positioning Concept

Many methods have been used to improve businesses. Improvement implies change (the reverse is not always true, of course); so improvement efforts have generally been called change projects in recent years. Every improvement method can therefore be called a change methodology. Re-engineering is certainly a change methodology. Information systems development projects may also be considered change projects, when they change the business. All these methodologies are used in special efforts, often called projects, but also called studies or interventions. In most cases, the change effort contains more activities than are covered by the methodology. For example, a change project typically includes the selection of a team, which is not usually thought of as part of the effort. These activities are usually considered common business practices, lumped into the topic of project management. When re-engineering was first done, it was managed by general project methods as a matter of course.

There is a growing realization that general project management (which covers planning, resource allocation, project tracking, and problem correction, among other activities) may not provide all the control required for change projects and for re-engineering in particular. The use of a broad change management framework would seem to be important for all forms of process re-engineering, which is itself a well-defined and narrowly focused activity.

This chapter describes the concept of managing change in the corporation using the Positioning approach of Dynamic Business Re-engineering. The implementation and used of Positioning is covered in Chap. 6. The goals of change, along with the relationship between change management and quality management, are also discussed in detail in this chapter.

75

Managing Change: Positioning

The change management approach used by Dynamic Business Re-engineering, called Positioning, provides a broad framework for controlling change in business. To do this, its scope encompasses all aspects of corporate change. Figure 4.1 shows the scope of each possible layer of change management:

1. *Positioning*: the framework of all corporate change.
2. Traditional project management methods.
3. Re-engineering: the change methodology.

The Positioning layer of this model represents the creation and use of a series of techniques, models, and concepts that form the basis for change support. This basis involves the definition of corporate business goals, the positioning of the company in the marketplace, and the positioning of the company to change rapidly in response to opportunity, market pressure, or regulation.

The objectives of Positioning also go beyond those customarily associated with change methodologies. These objectives are:

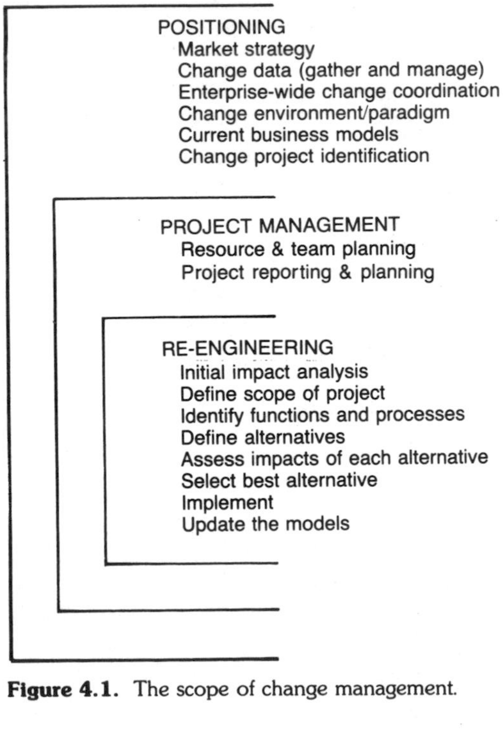

POSITIONING
Market strategy
Change data (gather and manage)
Enterprise-wide change coordination
Change environment/paradigm
Current business models
Change project identification

PROJECT MANAGEMENT
Resource & team planning
Project reporting & planning

RE-ENGINEERING
Initial impact analysis
Define scope of project
Identify functions and processes
Define alternatives
Assess impacts of each alternative
Select best alternative
Implement
Update the models

Figure 4.1. The scope of change management.

1. To control change systematically.
2. To make change as easy as possible.
3. To remove the threat associated with change.
4. To make change into a continuing series of incremental improvements.
5. To package change increments into projects of manageable size.
6. To organize corporate, departmental, strategic, and operational data related to change.
7. To gather data from market research, technology watch, and staff observations.
8. To coordinate quality programs across organizational lines.
9. To provide the environment and methods and for individual change projects, including re-engineering.
10. To manage change projects, and to evaluate the results of these projects when they are completed.
11. To determine the changes the company needs for competitive advantage.

Positioning is a basic reorientation in the attitude of the company toward change. It, not re-engineering, implements the change paradigm discussed in the last chapter. One reason that re-engineering itself does not alter either prevailing attitudes toward change, or prevailing abilities to initiate change, is that these shifts are outside the scope of the re-engineering process. Another reason is that re-engineering is targeted at specific processes, while the required shifts must occur throughout the company.

In effect, Positioning surrounds re-engineering. It determines key elements of market strategy, and positions the business to be re-engineered. After the newly re-engineered business processes have been implemented, Positioning provides a new attitude toward change. The mechanisms that transform many sorts of business changes into an easily used toolkit are provided within the methodologies of Positioning.

Goals of Change

Positioning starts with determining the goals for change. These goals may be broader than the ones used for re-engineering, which is much more focused. They may also be stated differently, due to the differences in character between positioning and re-engineering.

At a summary level, most re-engineering efforts have been undertaken for two reasons, neither of which have given the company much real choice: (1) to successfully compete and possibly gain a competitive advantage, and (2) to respond to some form of regulation or executive mandate. A third reason, which is found in an increasing number of cases due to the nature of business today, is to support a merger or an acquisition. We are finding that, as two companies merge, there are tremendous opportunities for re-engineering. In fact, a common problem is that of assimilating a newly purchased company. Only a few of the companies we have spoken to consider that they do a good job in this area, and others admit that they have nearly destroyed good companies because of poor assimilation.

The goals of change are the basic goals of business: increase profit by increasing revenue and decreasing costs. For this reason all of the basic objectives of business should be considered for every change effort. The goals of change specifically related to re-engineering efforts, however, are few and specific. The most common are to:

1. Streamline the operation.

2. Reduce costs.

3. Improve quality.

4. Increase revenue.

5. Improve customer orientation.

6. Merge acquired operations.

Each of these has been stated as the primary goal of one or more re-engineering projects. They are not, however, the only ones that are possible, and most can be divided. Some, in fact, imply others: streamlining and merging imply that the overall goal is cost reduction. Each of these goals is now discussed.

Streamline the Operation

Streamlining, the attempt to bring a business process to its most efficient form, is the stated goal of most re-engineering efforts. Although streamlining implies reducing costs and production (or service) time, the goal is not stated as cost reduction for several reasons. First, cost reduction carries a stronger connotation of workforce cutbacks, which may precipitate unnecessary trouble. Second, streamlining implies the desirable concern for quality and effectiveness that cost cutting may omit. Finally, streamlining most directly addresses the well-placed concern of management that old processes have become inefficient. The "lean and mean" approach goes beyond cost cutting, to be sure.

A streamlined operation provides efficient, flexible activity by eliminating redundant operations, improving work flow, improving support systems, and even anticipating the effect of actions on other departments. In a fully efficient operation, only that which is needed is done. Setup and waiting time in production areas is minimized. Work flow in operational areas is simple and direct, with all unnecessary tasks eliminated. Quality tends to improve, just because there are fewer places for things to go wrong.

The term "streamlining" may seem less than precise, but it is an excellent objective for re-engineering projects.

Reduce Costs

Cost reduction is one of the main reasons to re-engineer. While this is certainly a important goal, we recommend that it be viewed as a secondary objective, with operational streamlining and quality improvement being of more immediate concern. The reason is that, if these two goals are met, costs will be reduced. Furthermore, continuing progress toward quality and efficiency will continue to provide additional cost savings.

Our reason for making cost reduction a secondary goal is that too much emphasis is normally placed on this one factor. It often diminishes the attention paid to quality and effectiveness. It also often stresses immediate payback, and ignores the long-term good of the company. As re-engineering provides long-term improvement, stressing short-term savings is a philosophic (and practical) contradiction.

However, cost reduction is very often an inevitable objective of re-engineering. If an effective Positioning effort has preceded the re-engineering project, the sensitivity of the market to product pricing will have been analyzed, and target product costs will have been derived. Businesses often discover that they must set these externally mandated numbers as their goals, and then meet them, or fail in the marketplace. Cost reduction is even more compelling in the increasing number of cases in which re-engineering is undertaken to eliminate operating losses.

In addition to being a goal, cost reduction is a very effective performance measure, for both the change project and all future work. As always, potential changes must be viewed in terms of their costs and benefits. If re-engineering increases costs, which is rare but possible, a corresponding increase in benefit must be realized. The relationships will be clearly defined using the methods discussed in Chaps. 6 and 7, so that it will be possible to assess the impact of any action, with the resulting cost reductions or increases. It will also be easy to define the associated benefits and identify whether they are real or anticipated. With this information, better decisions regarding costs and actions can be made, and the success of a proposed change can be calculated in advance of implementing it.

While costs will be saved in all areas of the operation the most significant areas are labor, information, supply, administration, and the cost of money. In addition, it is sometimes possible to reduce the costs arising from taxes, tariffs, and fees.

Reducing the Cost of Labor.   The most common goal of cost reduction is conservation or outright cutting of staffing costs. If a process is made more efficient by re-engineering, it should need less effort. How else can the reduction of effort be realized than by reduced payroll costs?

Generally the lowering of payroll costs is equated to the elimination of jobs. This is the same approach to cost justification, and the same assumption regarding realizing benefits, that was used to justify new computer systems. Experience has shown that reducing payroll costs, which seems uncomplicated and desirable, has failed to provide long-term benefits. The overall staffing levels crept back up, requiring expensive recruitment, and the impact on the workforce was negative. They began to distrust technology. Not only was the cost reduction merely temporary, but hostility, frustration, and even occasional sabotage were created. In the final analysis, when one considers that automation eventually added jobs, the goal of reducing payroll costs was not the best one, and the implementation was worse. Re-engineering faces the same problems as these past technology efforts. The actual termination of existing employment should be weighed against all the costs and compared with all the many alternatives.

To reduce the overall number of positions in any organization, attrition and transfer/retraining are vastly preferable to involuntary separation. Attrition can reduce the payroll costs at a surprising rate. When people must be detached from the company, it should be clear to those who stay that the appropriate people left, ones whose skills and performance could not be used in the future, and that they were treated fairly. If the workforce, including middle management, is dealt with effectively, the fear of re-engineering will be replaced by trust, and the adversarial relationship between management and the working levels will be replaced by teamwork. The human capital side of re-engineering is at least as important as the business process restructuring. However, it is very difficult to abandon the goal of reducing staff costs.

Reducing the Cost of Information.   If information is considered an economic resource, like labor, capital, and raw materials, then reducing the cost of the information that goes into a product's development and manufacture can be considered a valid method by which costs can be reduced. Total product and service costs *do* contain information-related costs; these costs are already significant and can be expected to increase. These costs include design specifications for product types, ordering information for production lots, production methodologies, and materials requirements. While there is technologically advanced products (new drugs and military aircraft have hundreds of million dollars worth of information costs per type), even simple products and services are increasingly dependent on information for design, manufacture, and delivery.

Information costs also occur at the corporate level. Companies run on information. It is virtually their life blood (although many finance managers claim that the life blood of companies is money). Information supports every activity in every company. The sheer amount of information that any business deals with is staggering. It is also seldom adequately controlled or delivered, and so its costs are usually higher than they need to be.

Operational information costs include manual reporting, information handling (such as reviewing orders copying and filing documents), microfiche, memos, word processing, and computer-based systems. Every time a piece of paper or a keyboard is touched, an information cost is incurred. Viewed from the corporate perspective, information costs seem to be pure overhead, with little value added to the actual products. However, these costs can be allocated to products with full justification.

Information, as a significant cost element, is a ripe target for cost control. There is an optimum level of information cost for each product and for the enterprise as a whole. But most businesses are a long way from being in an optimal position in the use of information and the control of its costs. First, they must develop the ability to determine the optimum uses for information. Most companies will also need to upgrade their technology and communications architectures, and to invest in systems that support their long-range strategies.

Reducing the Costs of Materials and Supplies. The most common and successful re-engineering approach to reducing material costs is to reduce the amounts used. This can be done by reducing waste or by actual engineering changes. Another common concern is reducing inventory costs, which can be addressed through revised manufacturing processes and revised vendor relationships.

Reducing Administrative Costs. Often referred to as general and administrative (G&A), these are the costs of doing business that do not directly relate to the manufacture or delivery of products or services.

They are related to the physical plant, much of the information systems support, the communication support, accounting, purchasing, and, in short, anything that is not involved in production, marketing, or research and development. As such, this cost category offers an opportunity for considerable savings through streamlining and quality improvement.

Re-engineering efforts often focus on this area. Administrative support efforts have, in common with most corporate activities, evolved in the uncoordinated manner discussed in Chap. 2. Worse, administration does not normally receive the attention that operational work does. As a result, administrative work processes are usually inefficient. Also, layers of unnecessary complexity have often built up. Among the policies of almost every organization that is more than a few years old are rules and procedures that no one can justify. The term "bureaucracy" is most often applied to administrative work, both in institutions and in business.

Attempts to fix this problem in the past have often caused as much harm as good. Efforts have often failed to consider the complete context of the work being performed.

Successful attempts to re-engineer administrative processes have two important characteristics. First, they begin with the broadest view of corporate administrative processes. Second, they combine the analysis of administrative processes with operational processes, and set their priorities to achieve operational goals.

Costs of Money.   Most companies use a combination of internal funds and external debt for capital. Both methods incur money costs, in the form of interest or the loss of it. Also, shorter-term cost considerations associated with the time value of money must be applied to payments and collections.

The use of money is not often seen as a primary goal for re-engineering, but this element has considerable untapped potential. Lowering the cost of money can be addressed by re-engineering in two ways. First, the movement and costs of money can be included when studying operational and administrative processes. Second, it is theoretically possible to treat the cashflows as a business process and optimize them for the entire enterprise. While we are aware of several efforts to optimize cashflows, we have not seen any use re-engineering techniques.

Reducing Taxes, Tariffs, and Fees.   Reduction of these expenses is infrequently the primary goal of a re-engineering project. However, these costs can and should be minimized as a secondary goal in most change projects. While the savings to specialized types of businesses can be very significant, all companies could save in these areas. If any attention is paid to these costs, it is not difficult to reduce them by simply understanding the rules and building the most advantageous response to them.

Improve Quality

Like streamlining and cost reduction, quality improvement is nearly always a goal of business re-engineering projects. Improving quality in all processes will certainly increase the value of products and services, and reduce costs by cutting down waste. Quality improvement, as re-engineering goal, also avoids some of the resistance to re-engineering based on the assumed anticipation of workforce cuts.

Quality is also becoming an important, fundamental consideration in all business and institutional activities. This is most likely due to the degree to which it is credited with the success of Japanese export efforts. As a result of this emphasis, quality improvement programs worldwide are being given increased attention.

Quality is measured in terms of reliability, consistency, and longevity. The measurement of a product or service against these elements determines its worth.

Quality improvement for any process is directed toward improving the product and minimizing the amount of rework and scrap. It also involves the ability to obtain consistent results while adhering to increasingly high standards. The continual tightening of standards applying to all operational activities is indeed the best assurance that quality is improving.

Quality According to Deming

W. Edwards Deming is the person whom many credit with the Japanese industrial domination of the world. In the 1950s, Dr. Deming brought his concepts of production, quality control, cost savings, and continuous improvement to Japan. He found fertile ground for his ideas. The rest is history. Today, Deming can be viewed as the father of much of the change being made in the name of quality management in the United States. To compete with the Japanese, more and more companies are moving to adopt his concepts, hoping that they will do as much for them as they did for their competition in Japan. But the application of these concepts represents a significant paradigm shift for most managers and staff. A shift that requires flexibility and a willingness to challenge the status quo.

Much has been written about Deming's work, and his principles are widely quoted. However, we have also seen a wide range of interpretation of his ideas. Few interpreters emphasize the central concept of continual improvement. But this is critical to sound re-engineering, and a brief discussion of Deming's Principles (there are 14) is therefore worthwhile.

Deming's Quality Principles Applied to Re-engineering

Deming's principles have been proven to be effective when properly applied. Of all these concepts, the one with the greatest impact relates to the need for businesses to change continually to improve quality and to reduce waste and thus to reduce costs. But, again, in today's business operation, the application of these principles is difficult: companies are not geared for change.

The following discussion of Deming's principles, taken from his book *Out of the Crisis* (Cambridge, MA: MIT Center for Advanced Engineering Study, 1989), is an interpretation based on our application of these principles to re-engineering. The subject of quality improvement, and this interpretation in particular, is an important part of the Dynamic Business Re-engineering approach.

1. *Create constancy of purpose toward the improvement of product and service.* Commit the company to a long-term strategy for continuous change. Create a single corporate emphasis. Get everyone in the company behind the idea. Make everyone responsible for improvement. Make certain that each person knows what he or she must do, and the results that are expected.

2. *Management must take the leadership in promoting change.* Senior management must take an active role in re-engineering. They must not let these efforts be performed without their direct input (an approach some firms recommend in the name of empowerment).

   Senior management must also make a long-term commitment to change. The effort must be insulated against manager turnover, including that of the senior officers. To improve staff morale, the efforts must begin and remain highly visible.

3. *Stop dependence on inspection to achieve quality. Build quality into the product in the first place.* This is the foundation of total quality initiative (TQI) and continuous quality improvement (CQI). Quality cannot be achieved by checking for errors in the final steps of a process. It must be built into the product as a result of the work in the process.

   When re-engineering, concentrate on the processes, not on the organization. Build the checks and balances needed to ensure that quality is part of the work processes. Design quality assurance (inspection) into the processes to weed out bad product and as a feedback mechanism to improve the processes further. Also, use inspection as a gauge of ongoing success: the volume of bad products should decrease.

4. *Move to a single supplier for any one item. Create long-term relationships with suppliers.* Re-engineering can and should include and even the vendors own processes. This, of course, requires a close relationship with vendors and a very detailed knowledge of their practices and products.

   Re-engineering provides a good opportunity to consider reducing the number of vendors the company deals with. They can be culled by reviewing the quality of the materials they send or the services they provide.

5. *Improve constantly and forever the system of production and service, to improve quality and productivity, and thus constantly decrease costs.* This principle, as applied to re-engineering, is the basis of the Dynamic Business Re-engineering concept. As quality improves through continuous review and re-engineering, waste is reduced, redundancy is eliminated, and costs decrease.

6. *Institute training on the job.* As a company is re-engineered, it is important to work with the affected people and retrain those whose jobs will change. The more they know about their work and the process it is part of, the more likely they will be to find ways to improve it.

7. *Institute leadership. The aim of supervision should be to help people and machines to do a better job.* As processes and suppliers change in conjunction with re-engineering efforts, it is imperative that managers understand each phase of the new process. Re-engineering methods help by documenting the process thoroughly, and by illustrating the interactions between the process and the rest of the business. This will allow them to help the staff make the transition with the least possible amount of trauma. It will also help position the manager to evaluate ideas and to help develop proposals for presentation to higher-level management.

8. *Drive out fear, so that everyone may work effectively.* In approaching re-engineering, it is important that management clearly indicate its support for the workforce and assure everyone that few, if any, will lose their jobs as a result of the re-engineering. It is also important that management create an environment in which people are not afraid to present new ideas and try new approaches. The past fear of rejection and failure has cost many companies their creative edge. This must be reversed if true quality initiatives are to be implemented and if companies are to create an ability to respond quickly to marketplace pressures and opportunities.

9. *Break down the barriers between departments. Promote team building*

as people from different departments work together to solve problems and improve quality. Process re-engineering crosses departmental lines as the flow and activity of the processes are followed. For this reason, almost every re-engineering effort will provide significantly greater benefits if organizational lines are not allowed to become barriers. Also, the re-engineering change teams will normally be made up of members of different departments to address this problem.

This principle is difficult to put into practice without re-engineering. If quality circles or other interdepartmental approaches are used to improve quality, they will find their efforts frustrated by the inability to determine in advance which departments to involve and the inability to predict the effect of a proposed improvement across the entire enterprise.

10.   *Eliminate slogans, exhortations, and targets for the workforce asking for zero defects and new levels of productivity.* Zero defects and continually improving levels of productivity should be used as objectives for re-engineering and designed into new work processes. They must become the focus of all activity, and they must be installed as part of a new operational paradigm.

11a. *Eliminate work quotas; substitute leadership.* The production of a high volume of product that has a high rate of error is obviously worthless and expensive. Re-engineering should instead concentrate on a balanced set of production objectives that stress high quality and consider volume as a secondary concern.

This is a key concept in re-engineering. The process improvements will provide little or no benefit unless quality is emphasized over volume.

11b. *Eliminate management by objective; substitute leadership.* Management by objective (MBO) is not incompatible with re-engineering, in that the concept of a contract between manager and subordinate can be made. However, re-engineering and total quality management both do best when the management relationships are more open and broadly applied. The objective is to improve, not to meet an arbitrary target.

12.   *Remove barriers that rob managers, engineers, and the hourly worker of their right to pride of workmanship. Change the emphasis from numbers to quality.* Almost everyone wants to do a good job. This motivation comes from the universal need to belong to the group and to feel important. Most people also take considerable pride in their company and in what they make or do. To a large extent, when

these feelings are suppressed or are absent, the amount and quality of work suffer. Also, labor relations problems increase as management tries to compensate for lower levels of productivity by imposing quotas and blaming staff for poor quality.

Re-engineering and the move to operate in an environment that promotes participation in improvement by everyone in the operation provide an opportunity to begin the process of promoting pride and a sense of self-worth. Both are important factors in the long-term success of a company.

13. *Institute a vigorous program of education and self-improvement.* This principle is outside of the scope of re-engineering, having more direct implications for general management. However, re-engineering does provide the means to implement the improved individual capabilities and the technical advances that both management and staff will discover if education and self-improvement are effective.

14. *Put everyone in the company to work to accomplish the transformation. Make it an all-pervasive common goal and support it.* The type of transformation that is possible through the change paradigm requires the involvement of everyone in the company. No one can be left out. However, this involvement will be gradual. Not everyone will be involved at the start of the project.

These principles have much in common with those of the change paradigm and the objectives of Dynamic Business Re-engineering's Positioning methods. Deming's views and ours coincide on the need to effect change as a series continual improvements. However, we believe that quality programs that do not make use of Positioning, or some equivalent corporate-wide umbrella-type change in the management environment, will be unnecessarily restricted. The restrictions that Deming-inspired quality improvement programs face is that they lack the ability to assess the impact of their ideas in advance and to coordinate efforts across organizational lines. These restrictions can be removed through the use of Positioning models to support operational simulation and impact analysis. The most effective approach to quality improvement is, therefore, a combination of TQM and re-engineering, under the control of Positioning.

Increase Revenue

Obviously increasing revenue is, with cost reduction, one of the two basic methods by which profit can be increased. Revenue can be increased either by increasing the price of each product or service, or by increasing the number of products (or the amount of service) sold. Increasing price usually causes a decrease in sales volume and in any case does not require any changes in the way business is conducted. The associated impact on volume or on market share is, however, a valid re-engineering concern.

Increasing revenue is not stated as the principal motivation for re-engineering as often as might be expected. In most of the cases where this is the principal motivation for re-engineering, a completely new product is involved.

One of the reasons for the apparent lack of interest in this goal is that it implies an unavoidable risk. The re-engineering investment must first result in the achievement of the project's internal goals, which are under the control of corporate management. The marketplace must then respond by supplying increases in revenue, which is not a controllable factor. As a result, even if management has external goals in mind, they are more likely to pose internal ones for the project.

Another reason for eschewing this objective is that the easiest way to increase revenue is to decrease costs, which is usually stated as the goal of the project. In fact, the most obvious changes to the business process that may increase sales (reducing costs, improving quality, and cutting production time) are usually stated as goals themselves. However, there are other ways that re-engineering can address increasing revenue. For example, new or changed products and services can be offered; re-engineering is especially well suited to the development of new services. Other revenue-increasing measures can include the improved collection of receivables, and shortened product development, production, and delivery times. In this way it is possible to increase the effect of a revenue stream without increasing its peak amounts.

Quality improvements can also have a net affect of increasing revenues. Quality improvement directly contributes to the value and marketability of a product. In addition, improved quality throughout a process decreases waste, increases production capacity, and decreases delays, which can increase the effect of revenue as well.

This identification of opportunities to increase revenue is a key by-product of the market and quality analysis performed as part of Positioning. Given its importance, as Positioning becomes more common, this goal may become the most prevalent objective for corporate change projects.

Improve Customer Orientation

Improving the orientation of the company toward its customers has been stated as a significant re-engineering goal in a high proportion of recent projects. Since the customer's assessment of the company and its products is strongly influenced by service, improvements in this area are another good way to increase revenue. A strong customer orientation is especially important for service industries, the service aspects of manufacturing and for not-for-profit institutions.

Merge Acquired Operations

Corporate mergers and acquisitions are increasing in number and in importance. When a merger takes place, there is a high probability that an attempt will be made to reduce redundant effort by combining functions. This can be only done by re-engineering methods, and there is an increasing propensity to do so.

An interesting example of this approach has been undertaken in the People's Republic of China. The government decided that China should become a competitor in the consumer electronics exporting business. A new company, called Chinatron, was created and five existing companies were earmarked to be combined to form the new company. Re-engineering techniques were selected to be used to combine the five companies and transform their current operations, which had little to do with the production or export of retail electronics, into a new business. This is probably the most ambitious single re-engineering project ever attempted in terms of its scope and complexity. The effort shows considerable promise.

Changing Goals: Positioning

If re-engineering is the best way to make business processes achieve the goals set for them, then Positioning is the best way to set and to reset these goals. But Positioning itself has goals, which are best pursued during the early stages of implementing Positioning in the company.

The Goals of Positioning

The goals of positioning include setting marketing strategy, setting the environment for change (especially re-engineering change), and determining the details of the company's current operation. The activities required to implement Positioning are diverse, but they are interrelated by having common goals and common timing and by involving the organization's most senior managers. The first activities of Positioning will be to gather information about the company's goals, plans, and strategies.

Market strategy setting, an integral part of Positioning, is based on the corporate review of both its markets (by product line) and its competition's capabilities, strengths, and weaknesses. The opportunities and actions required to eliminate internal weaknesses and capitalize on competitor's weaknesses are analyzed against the baseline information of the Positioning models to determine effort and impact. Setting a marketing strategy involves broad-based decisions regarding the market positions to which both the company and its products should aspire. When applicable, this strategy will also include decisions related to the development of new products.

Corporate market positioning decisions today may have a newly broadened scope due to globalization, and include consideration of such factors as global competition, international partnerships and the development of international assets. To ensure the proper focus, all corporate change projects should have immediate access to the most current corporate market strategy and the individual market plans for each product and service. At the lower echelons of the change projects, this type of information should also include details of what the company says in its advertising, what the competition is doing, and what the customers' attitudes are.

The most disciplined part of Positioning is the gathering of data. The types of information gathered by Positioning and used for change projects are shown in Fig. 4.2. This is necessarily a detailed process, and will normally take some time, especially for very large and old operations. However, the time and effort can be reduced and the effect of the data increased by using automated tools. Dynamic Business Re-engineering is supported by the Positioning and Re-engineering (PAR) system, which runs on personal computers and is designed specifically for this purpose. However, any computerized system that will keep text, such as a word processor, and draw charts is better than using only paper. Once gathered, this data must be kept up-to-date. Furthermore, the results of this activity are important not only as input to the re-engineering effort. Through creative application, this information will have many uses in supporting daily activity and decision making. This use will, however, require flexibility and computer-based support.

The collected information should address each facet of the operation. It should include work flow models (BAMs), organization charts, mission statements, business process models, business policy and rule statements, job models (relational diagrams), and corporate plans. This information contains answers to the interrogatories of who, what, when, where, how, and why. Together, this information forms the Positioning baseline definition of the company.

All change is based on an analysis and modification of this information. As such, this baseline provides the foundation for operating within the change paradigm and for all re-engineering activity.

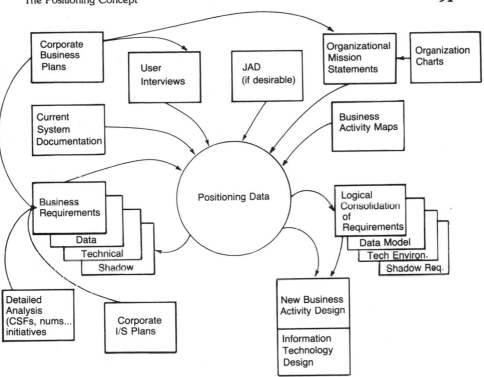

Figure 4.2. Positioning data.

The Change Environment

The second goal of the positioning effort is the creation of an environment that is amenable to change. This must be done before plans are made for re-engineering, indeed before the overall corporate strategy is laid out. The most important aspect of establishing a change environment is a new attitude toward change, of course. There is unfortunately no simple plan for changing attitudes. In addition to a new attitude, the change environment has a few simple physical components.

The most visible element of the change environment is a corporate change infrastructure. This can be a standing Positioning staff, headed by a chief change officer (CCO), with the authority to carry out positioning activities throughout the company. Alternately, the infrastructure can be the management of the company itself. The latter approach will be effective only if the chief executive officer of the company becomes the chief change officer, and adds the management of positioning to his or her daily activities. In fact, the company's chief executive will be a de facto chief change officer, even though the job is not wanted or even recognized, if no other executive is given this responsibility. Whether or not there is a CCO, Positioning requires staff support, and this effort must be connected to management in every department.

Given the organization to carry out Positioning, the other key element is the support system needed to maintain Positioning information and to provide access to the models of the business's processes. These models will be used to simulate the proposed changes and to assess their effect on the operation. All of the activities related to information collection and use are candidates for automation.

By building the Positioning environment, the business will be laying the groundwork for change projects. This information not only facilitates these efforts, it also shortens their length by eliminating the need to define baseline information for the operation each time a change project is undertaken.

Where We Are Today in Business Re-engineering

A great deal has been written in the early 1990s about re-engineering projects. Many efforts have succeeded in toto, some only within limited parameters. A few have failed. Most of the projects that have failed have been abandoned, rather than failing to yield the minimum results expected after completion. It is unfortunately common to attempt to use re-engineering techniques to resolve problems that need to show results in a few weeks and to discard the process when it is discovered that months will be required. Also, a handful of re-engineering projects have failed after implemention and have been backed out of production. The manager's nightmare—having a re-engineering project ruin a company—never seems to have happened.

Successful Re-engineering Work

There are many good examples of recent successful re-engineering work. One of the more spectacular is the complete re-engineering of McDonnell Douglas's product development cycle. When the aircraft manufacturer realized that its pricing for new military types was not competitive, the company decided to cut 40 percent from the development cycle costs, from concept through production. Reviewing existing processes, it was noted that each new aircraft was designed several times, first when the initial concept was laid out, then when the prototype was built, and again for production. In addition, producing the maintenance manual could be viewed as another design effort. Maintenance specifications for military aircraft are very expensive; it is said that the manual for the B1 Bomber is over one million pages long. Each of these processes was somewhat independent of the others. The suggestion was made that the design, which had been computer-assisted for many years, be done once and then used from the same computer source for all of the processes thereafter, with appropriate additions and modifications, of course.

The idea was implemented with great success. The restructuring of technology that was used (another McDonnell Douglas specialty), as well as some additional technology. The improvement was considered motivated by cost reduction, but its goal was more often described as streamlining. The real savings was in the cost of information in each new aircraft type, by far the largest expense involved. This project, which was not fundamentally thought of as a re-engineering project, was none the less a fine example of re-engineering using information technology to reduce the cost of information in products and thus lower product costs significantly.

Northwest Airlines is another example of the use of information technology to enable streamlining of business operations. One of the most difficult tasks that an airline is required to perform is postprocessing of tickets. Tickets come in several forms, and are processed in many ways. They move everywhere, and there is no guarantee that they will be in good condition when they come back to the airline. The charges and adjustments made to them are also extremely complex. As a result, tickets have always required a significant amount of manual handling, which is, of course, very expensive for the airlines. Northwest applied a very new technology, image processing, to address this problem. Using optical scanning, and placing the exact image of the ticket on optical disk storage, the tickets could be processed by computer to a much greater extent. Until recently, the large amount of computer storage space required to store images made this approach economically infeasible. However, implementing the new technology itself did not save any expense; it was restructuring the work that saved effort and reduced the time required to collect interairline fee transfers.

What Has Been Learned?

From the projects that have re-engineered businesses, some fairly clear lessons have been learned. The first and most important is that it can be done. Re-engineering can accomplish objectives set for it, frequently very ambitious ones.

Re-engineering cannot, however, be done halfway. Once the decision has been made to re-engineer, the only sensible limitation that can be applied is to the scope of the effort, and even the scope should not be too constrained. The other most consistent qualification is the direct involvement of executive management, from start to finish.

The Dynamic Business Re-engineering Approach

Dynamic Business Re-engineering is a combination of Positioning and re-engineering using some particular methodologies for each. There are other methods for re-engineering, but none addresses the range of activities covered by both re-engineering and Positioning.

In addition to having a wider scope, Dynamic Business Re-engineering differs from the other approaches in several important ways. First, and most important, is that Dynamic Business Re-engineering creates and supports a new dynamic model of business process rather than just restructuring the static model. The second difference lies in the use of relational systems development (RSD), a proven technique, as a basis for business process modeling. RSD is itself unique in that it begins by modeling the existing business. A third difference is that Dynamic Business Re-engineering directly addresses business processes, distinguishing between business process, work flow and organization. Finally, Dynamic Business Re-engineering does not discard its models after the project is completed. A consequence of its dynamic nature, the models produced by

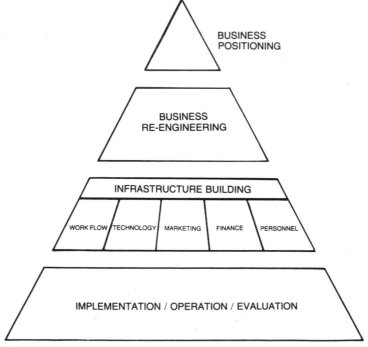

Figure 4.3. Dynamic Business Re-Engineering.

each re-engineering project become operating tools and inputs to new re-engineering efforts.

In some cases the design of the new business process structure can be simple and therefore informally described. The most successful of re-engineering projects have been able to get by with little complex process design and almost no modeling. Most future efforts will not be as fortunate; they will need powerful, but easily used, tools to help with these functions. Additionally, retaining the process models permanently, so that they can be used to monitor performance and continue to improve the business on an ongoing basis, reinforces the need for effective modeling tools. For most purposes, it will be possible to control complex change projects only with a combination of effective methods and easily used automated tools.

Dynamic Business Re-engineering uses two sets of automated tools: 1) a set of positioning tools that gather data about the current environment and the goals for re-engineering, and (2) a set of tools that automate the RSD methodology, providing the modeling capability for re-engineering. The Positioning tools take an accurate picture of the current operation and gather the plans of the company, the results of the Positioning activities related to marketing, and the re-engineering goals that are thought to be most appropriate. The re-engineering tools use this information to create models of the business processes and to simulate the changes that are proposed. Because RSD was first designed to aid computer systems work, the models developed using RSD to support re-engineering can also be used to support the creation of any ancillary computer systems.

The re-engineering models that Dynamic Business Re-engineering uses allow management to try new process designs before they are implemented. Each business function can be studied, experimented with, and modified by corporate managers. The results are a well-designed new business process, a clear picture of the impact of the changes and a well developed knowledge of the processes.

Summary: What Is Beyond Quality and Efficiency Gains?

Even the most advanced thinking concerning quality initiatives assumes that, if there is enough quality improvement, success will follow. But quality gains without corresponding efficiency increases can slow the production process and increase labor cost. If this happens, nothing will be gained. The same holds for efforts that focus only on efficiency. If all processes are made as efficient and thus cost-effective as possible, will the product be marketable? Will quality be maintained? Other management and business change theories follow this pattern, but most managers will not trust the futures of their businesses to them.

So cutting cost and improving quality are both useful only as far as they go. Quality improvement does provide the happy combination of reduced costs and increased product worth. But these two objectives will work with complete reliability only when they are applied within a complete management framework; a framework that understands and considers the relationships between actions.

Positioning is a complete management framework for business and other organizations. It helps management discover where the organization should be and lays the groundwork for getting there. Quality and efficiency gains are within its scope, as deliberate, managed objectives for every business process. It is indeed necessary to go beyond quality and efficiency improvements to operate in a world of fierce competition.

5

The Tools Used
in Modeling
the Business

The Dynamic Business Re-engineering methodology represents an expansion of the relational systems development (RSD) methodology. RSD was developed to integrate business operations and computer systems support. The RSD approach recognized the need to begin with an understanding of how a business operates and then determine how automated support can improve the operation's effectiveness, while providing consistency of work outcomes. RSD views support from the business operation inward to the support services. It is business oriented, and not information systems oriented. The result is that systems are built to solve the needs of the business as defined by the operation.

As a consequence of the RSD approach, the business operation work flow was changed while the new information system was being designed. It was streamlined, and the quality of work was improved. The result was a new business design that had appropriate supporting computer systems. This created an integrated operational design in which the business processes and the computer support become very tightly coupled.

But this still failed to go far enough in supporting the restructuring of the operation. There were too many problems unaddressed by the methodology. Key among these was that change is constant in a business, and that the operations and information systems never seem to be able to adequately adjust. Also, the tools of computer systems development were really created for a one-time use. They are not easy to modify or keep up-to-date.

Furthermore, while the application of the traditional management sciences provided immediate relief, companies would soon find themselves in the same position as before. Business simply has not supported the operational models on a continuing basis. There has generally been little value perceived in doing so, but the need to respond to change is as important to re-engineering as it was to information systems development.

This chapter describes the basic forms of the tools that have been used to support various types of change methodologies in modeling business processes. The ones recommended for use in re-engineering, particularly Dynamic Business Re-engineering, are discussed in greater detail.

What Is a Business Process Model?

A *business process model* may be defined as a representation of the company's operation or a specific part of the operation. It is usually a graphical depiction of the structure and activities of the operation. The model often shows the relationships between work steps and their sequence. Together, these representations portray work flow.

The corporate model is composed of many individual interrelated models. These models will differ based on the areas they address and the modeling techniques used. In Dynamic Business Re-engineering the primary tools are the Business Activity Map and the Relational Diagram, although other modeling techniques (like organization charting and specialized techniques such as systems interface maps depicting the interaction of computer systems) are also used as necessary.

Regardless of the techniques used in the individual models, all the detailed models form an integrated whole. For example, as the flow of an activity leaves one department, it must pick up again in one or more other departments. The ability to track the flow across these boundaries is critical. When this can be done, the models are integrated. To accomplish this synergy among models, it is necessary to set standards that direct the manner in which the models will link together.

Typically, a model contains information about each work step and about each aspect of the operation's performance and support. For example, for a company with several locations, the model tells what each location does, when it does it, why it does it, and how the action is performed. In addition, the supporting information will discuss the information services support, all applicable business rules, and the interaction with other work steps, work flows, and processes. The relationships between the various locations are also shown, as partially assembled products are sent to other locations for the next step.

To be complete, the model must show all activity and the relationships between:

Each department's mission and the activity the department performs.

Activities (work flow).

Activities and process.

Rules and process.

The department's plan and its processes.

Activities and jobs.

Through its supporting information, it must also answer the questions of who, what, when, where, how, and why for each activity. And it should describe all support for each activity. Any outsourced activity should also be described along with its requirements. Examples are the use of external credit ranking and credit collection agencies.

This chapter discusses the tools used to model and re-engineer business operations. An emphasis has been placed on Business Activity Mapping and Relational Diagramming, two of the primary tools of the Dynamic Business Re-engineering method.

The creation of the Positioning models is discussed in Chap. 6. This discussion addresses the application of the tools and the results of their use. The development of process models for use in re-engineering is addressed in Chap. 7. The two main factors that motivate change are also discussed in that chapter, as well as the use of the models in simulating proposed new business processes.

Tools Used in Modeling a Business Operation

The tools about to be discussed are used to create business process models. Each obviously has its strengths and its weaknesses. We believe that the best features of all the others have been incorporated in the Business Activity Map and the Relational Diagram. However, other re-engineering consultants may use or recommend any of these tools.

Tools Used to Represent Business Processes

Business process modeling is nothing new. Management, industrial, and operations engineers have attempted to model activity for many years. The tools used have varied, but all have had the same purpose: to depict flow and work steps. As different companies have attempted to improve their operations, many different techniques have evolved, including the following ones.

Flowcharting.   Flowcharting, one of the oldest forms of work flow modeling, is a graphical representation of the sequence of steps in a task or activity. All activity flow is described by discrete symbols, and the symbol for each step is drawn with respect to the ones immediately preceding and following it. Flow is typically shown as a line with a directional arrowhead. The initial step in flowcharting is to define the work steps and their sequence. All decisions and relationships are next identified. Finally, the work steps are drawn in a straight-line representation of the flow. Decisions are shown as branching, where one or another choice is made. Figure 5.1 shows an example of flowcharting.

Tree Diagram.   In this traditional decomposition technique, a breakdown is shown as limbs from a central tree trunk. As one goes farther down a limb, it branches to show how something splits into component parts. This branching continues until the desired level of detail is shown. Using this technique, an operation can be repeatedly divided until all tasks are identified. While this technique is useful for demonstrating the

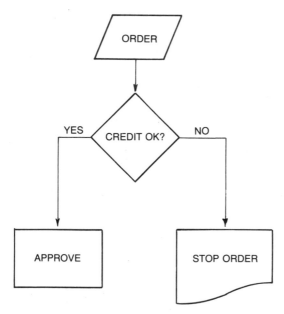

Figure 5.1. Flowchart.

breakdown of functions, it does not show flow. A type of tree diagram is the decision tree, showing the decision logic of a process, which is to some extent able to show business rules.

Warnier-Orr Diagram. Warnier-Orr diagrams are decomposition charts, specifically showing the hierarchical structure of business functions or systems. They are laid out horizontally, rather than vertically, but otherwise they are not very different from tree diagrams. Figure 5.2 shows an example of this type of chart.

State Transition Diagrams. To provide logic for digital processes, a diagram showing processes as a connected network of discrete states is sometimes useful. For this purpose, states must be defined for each station in a work process, the two simplest being active and waiting. The rules governing movement from one station to the others are then defined, and from that point it is easy to design a computer program or simulation model. State transition diagrams are most useful in re-engineering when the process being re-engineered is already heavily automated. Otherwise, the states may be difficult to define and artificial. Unfortunately, because these diagrams are also not easy to understand, they are poor tools for the managers involved in re-engineering.

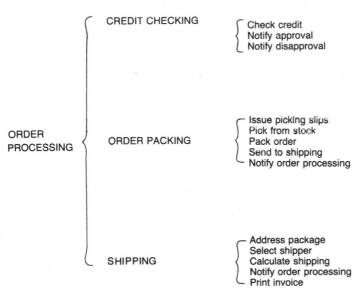

Figure 5.2. Warnier-Orr diagram.

Fishbone Diagrams.   Like tree diagrams, this technique operates using a central process line. Major activities are placed along this line as angled intersecting lines. The placement of these activities represents sequence, starting at the front or right side of the model and moving left. Activity components are noted along these angled lines on smaller lines running parallel to the main process line. This breakdown continues and can reach multiple levels. While this model shows basic relationships and the components of an activity, it does not show flow and only shows sequence at a general level.

Hierarchy Charts.   These models are decomposition diagrams, similar to trees. Beginning with a single global statement of action, they are then broken down (or "factored" or "decomposed") into lower levels of detail. The relationship is vertical, showing how actions at each level are divided into components. Theoretically, every action at a given level is at the same level of detail, but this is seldom true. This technique provides breakdown, but again does not show flow. An example of this technique is the standard organization chart.

Synaptic Models.   Synaptic models attempt to simulate the workings of the human brain. They represent a matrix or network view of activity. Processes are paths through the model, with the nodes being either activities or organizational units, but not both.

Network Models.   Network models begin with a single starting point and then show each successive step. Flow is implied by the position of the step. The relationships between the steps is shown by their placement. Timing and iterative activity are often difficult to depict in this tool. A commonly used application of a network model is the PERT chart.

Computerized Simulation Models.   A simulation model artificially reproduces the behavior of a real process. These models are generally computer programs that can be relied on to demonstrate the changed behavior of a process when key variables are altered. An example is an inventory model, which shows outputs such as inventory levels and costs, given the input of stock withdrawals and replenishments. Some of these programs are built using special simulation languages, that allow the programmer to structure such business arrangements as service facilities, shop floors, and assembly lines. Simulation modeling can be used to support re-engineering efforts when numerical analysis is required of the alternatives being studied. Dynamic Business Re-engineering uses the data gathered to support its mapping of the business as very generalized, undetailed simulation models.

Mathematical Models.   Mathematical solutions, called mathematical models, are applicable for certain business problems. These solutions are models in the sense that the business uses its real data as input to them. Although there are several forms of mathematical models, the most useful is linear programming. Linear programs are a series of simultaneous linear equations, called *constraints*, and what is known as an *objective function*, which is another equation that tells the model what the business wants to optimize. For example, a linear programming model may be used to determine the best mixture of ingredients for a cosmetic. The objective function would be the sum of all of the costs of each ingredient, expressed as the amount used multiplied by the unit price of each, and the constraints would be the upper and lower limits of each ingredient that must be used to make the product. Mathematical programs are used for specialized problems in business and can seldom be of use in the overall design of a process.

Action Work Flow Models.   A very new approach to representing business processes is the development of a new language. Terry Winograd, of Stanford University, and the Action Technologies Company, of Alameda, California, are collaborating on a work flow management technology that has the potential of becoming a breakthrough in most basic work flow management methods and perhaps in the management of work in general. The addition made by this group is that the interactions among the members of a group at work, which have been overlooked or assumed to be a constant factor, are being defined and supported by automation. This approach is most beneficial in service work flows, for which the customer can now become part of the model, and for knowledge work, which has generally been left out of work flows because it did not seem to be process oriented. Since these types of work are increasing, the new approaches being developed by Winograd and his colleagues would seem to offer a great potential advantage to re-engineering.

RSD Business Activity Maps (BAMs).   BAMs are flow diagrams that identify the activities being performed and depict the flow of work and the relationship between activities. It shows all decisions and the branching flow paths that result. All logic and rules are cross-referenced through comments imbedded in the diagram. The antecedents of this type of diagram have been called bubble charts, data flow diagrams, and work flow diagrams.

RSD Relational Diagrams.   Relational Diagrams are used to model how a job is performed. This is the first technique that depicts the interaction between a person's activity and the systems or operations supporting it. This interaction is shown as a flow that moves from action to action. As work is performed, the relationships between actions are described. As an action is supported in some way by a computer system or external service (credit approval process or the like) the work flow moves from that of a person to that of the support action. All associated information and logic are also shown in this technique. This model uses flow charting techniques to describe flow.

Business Activity Maps and Relational Diagrams

Both Business Activity Maps and Relational Diagrams are central elements in the relational systems development methodology, and a more detailed discussion of each can be found in our book titled *Relational Systems Development* (New York: McGraw-Hill, 1989). In general terms, BAMs are used to identify the actions that are being performed and show their flow and relationships. Relational Diagrams pick up after the lowest level BAM and are used to depict how the actions are performed.

Business Activity Mapping

Business Activity Maps (BAMs) are the primary technique used to create work flow models. Through the use of this technique, all operational activities will be identified and defined. All relationships to other functions will also be noted as interface reference points, and all flow will be modeled.

The purpose of BAMs is to provide comprehensive models of work activity flow and work process flow. They provide all the information needed to understand a business operation through both a graphical representation of work flow and associated detail information.

Unlike similar presentations in information systems, BAMs do not show data or information flow. The information or data used in the action being depicted is considered to be one of the pieces of descriptive information about each BAM. As such, it is not emphasized and is of no more consequence than any other supporting information, such as the business rules that direct the operation.

BAMs are used at four points in the Dynamic Business Re-engineering method. First, they are used in the Positioning setup to describe the current work flow and later, once all business functions have been identified, to reconstruct work processes. The third use is in re-engineering where they support work flow simulation modeling. Finally, they are used to implement the re-engineered operation.

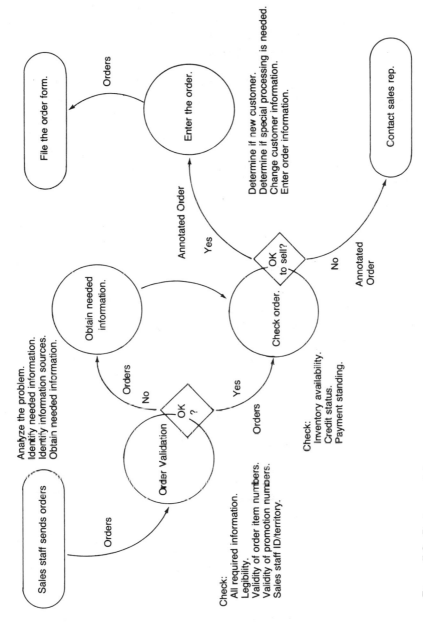

Sales staff sends orders

Orders

Check:
All required information.
Legibility.
Validity of order item numbers.
Validity of promotion numbers.
Sales staff ID/territory.

Order Validation

OK ?

No

Orders

Yes

Orders

Analyze the problem.
Identify needed information.
Identify information sources.
Obtain needed information.

Obtain needed information.

Check order.

Check:
Inventory availability.
Credit status.
Payment standing.

OK to sell?

Yes

Annotated Order

No

Annotated Order

Enter the order.

Orders

File the order form.

Determine if new customer.
Determine if special processing is needed.
Change customer information.
Enter order information.

Contact sales rep.

Figure 5.3. Business activity map.

105

By nature, BAMs are both networked and hierarchical. The initial flow effort begins with the question, what is the department responsible for—what do you do? The answer to this question will be a list of activities. Depending on the complexity of the activity, it can be broken into many lower levels of detail. The most complex activity we have encountered required seven separate levels of decomposition. The norm, however, is three or four levels.

In process decomposition, there are no definitive guidelines as to how many levels are appropriate for a given situation. There are also no rules on the content of any level. The reason is that the number of levels and the content of any level are irrelevant. Many approaches to decomposition disagree with this point and require analysts to make certain that all entries at any given level are at the same level of detail. But the goal of decomposition is to move the analyst or manager from the highest level of detail encountered to the lowest: the business function level. The intermediate levels are strictly to help break activity into lower levels of detail in an organized manner. Also, in practice we have found that, in any activity, the complexity of each task and thus the number of decomposition levels necessary to reach the business function level will vary.

Business functions are defined as groupings of tasks that perform a specific action or produce a specific end result. In Dynamic Business Re-engineering, this definition is further refined. This business function level is reached when the analyst stops looking at what is happening and begins to look at how it is being done. Business functions are boundary-specific, in that their tasks form a discrete work unit.

When the business function level of the BAM has been drawn, all information that is relevant to the work within the function is defined and associated with the diagram. All interactions with other functions and all timing data are entered and cross-referenced to the function's graphical flow representation. This completes the BAM.

The Analytical Attitude

When gathering data for the BAM; the analyst must refrain from criticizing or, as happens surprisingly often, deriding what is being done by the business operation staff. Their function is to collect accurate information: anything that is left out of the BAM may be left out of the re-engineered process. They must be open to diverse ways of doing business, and they must not color what they learn with their personal paradigms. All information in the BAM must be factual, not interpretive. The analyst must keep pursuing the same question until the activity is understood from the viewpoint of the workforce. Even terminology substitutions and inferences are dangerous. For these reasons, it is important that the analyst have a high degree of flexibility and tolerance.

In addition, the policies and rules for the process have a significant role in defining its paradigm. It is therefore important to identify the policies that apply to the process, and to obtain several interpretations of each policy's application. So much is left to individual discretion in the use of policies and rules that only consensus can guarantee accuracy. Also, if a policy is to be changed, knowing its actual use will be very valuable in planning the change.

BAM Components

Business Activity Maps are comprised of a number of symbols that represent specific operations. For example, BAMs use action symbols, decision symbols, flow initiation/termination symbols, flow connection lines, report use symbols, off-page connection symbols, and external BAM connection symbols. These symbols, shown in Fig. 5.4, are a core group, to

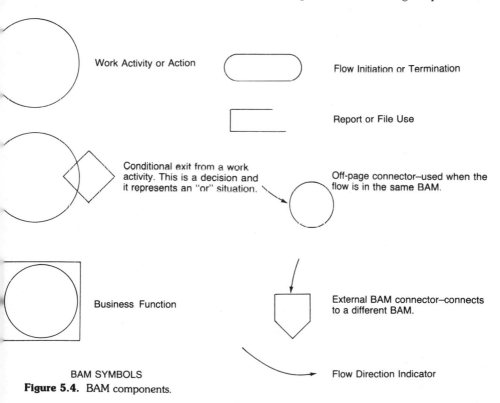

BAM SYMBOLS

Figure 5.4. BAM components.

which others may be added. Any addition should, however, be governed by corporate BAM standards to ensure consistency.

The symbols presented in Fig. 5.4 are standard throughout the U.S. business community. They are found on most drawing templates, and they are available in most automated drawing tools.

Action Symbol.   BAMs are comprised of a series of circles, called bubbles, which represent actions. Each bubble represents a separate work step. A brief identifying name is given to each bubble, along with a number (see "BAM Numbering," page 110).

Each bubble can be connected to other bubbles. It is common for a bubble to have two or more exits. A bubble may also have conditional exits. A single exit represents a linear flow from one bubble to the next: do A, then do B. Multiple exits indicate an "and" condition: do A "and" do B "and" do C "and" do D. Conditional exits are related to decisions, representing an "or" condition: do A "or" do B "or" do C.

When the business function level of decomposition is reached, the action bubble is placed in a square. This symbol indicates that the action is at the lowest level of decomposition.

Decision Symbols.   Many work actions will consist of both decisions. These decisions support conditional responses and result in alternative exits from the action: the selection of the next bubble depends on the result of the decision. If the action incorporates multiple decisions, either they can be grouped, or the action bubble can be split into more detailed bubbles. The decision symbol, a bubble with a diamond touching it, is used to represent this condition. The placement of the diamond is a matter of convenience only. Two or more flow lines can be drawn from the diamond, depending on the number of conditional options in the decision. Each flow line must be clearly labeled with:

1. The name of any documents being passed from one action to another.

2. The description of any other data that moves between bubbles.

3. The condition or decision alternative it represents.

Flow Initiation/Termination Symbol.   Each flow is initiated or stopped by an oval symbol. Initiation is always event related. A form is passed to an activity, an order is received, and so on. A flow is terminated when a specific leg of the flow is completed. Completion is usually accompanied by a single action such as "file the document" or "send the customer something." Each start or stop represents a boundary. Boundaries can be organizational, external (outside the company), internal, or activity related. Within a flow, each bubble's activity is initiated by whatever is passed from the prior bubble.

When the flow is interrupted while waiting for something to be returned, the action temporarily exits the flow. Later, once the item is returned, the flow resumes. The flow may, however, resume at a different point depending on external conditions. For this reason it is important to place a note describing the time-related nature of the exit at the exit point and at all re-entry points. Each of these exit and re-entry points should be cross-referenced. A start/stop symbol (oval) is normally used to depict exit and re-entry points.

Flow Connection Symbol. Each bubble is connected to one bubbles or to a flow start or stop symbol via flow connection lines. Arrowheads show the direction of the flow. Each flow connector is labeled with the name of the document or other item being passed. This document may be a product of the previous bubble, or it may be carried from its origin through several actions (bubbles). Each document that is passed at any point in the flow should be described in the supporting information. At a minimum, this information should contain a description of the document, its point of origin, and its purpose.

The use of this symbol indicates that the flow always moves through this path; it is not conditional. When both flow connector lines and decision-based flow are used in the same bubble, the flow will always continue through the connector lines, regardless of the decision-based flow continuation. For example, where A and B are flow connection lines and C and D are connections leaving a decision continuation (see Fig. 5.4), the flow always leaves the action bubble at both A and B and may also leave at either C or D.

Report Use Symbol. Many actions require information from reports or other documents. This information requirement is shown through the use of the report use symbol, a rectangle with an open side. This symbol is also used to indicate the filing of information in either a manual or automated file. In the automated system case, the symbol represents the computer system and its data file. This symbol is placed outside the action bubble and is connected to it by a flow connection line. The report name is used to label the flow connection line, and the place of the report's retention is used to label the report use symbol. In the case of an automated system, the system and file name are used to label the report use symbol.

Off-Page Connection Symbol. In many cases, several pages will be needed to show the flow of an action. As the flow moves from page to page, an off-page connector will be needed to show how the bubbles of one page connect to those of another. This symbol will be labeled with the page number, the bubble number, and the names of the bubbles it is connected to.

Automated drawing tools may require the division of a BAM into pages to allow the diagrams to be printed. If this occurs, care must be taken to add off-page connectors when dividing the tool's drawing board into pages. Even though the tool does not require them, they are necessary to follow the flow on printed pages. When using a computer screen, the automated tools normally allow one to work with several pages at a time by moving the view in or out and from side to side. Connectors are thus not needed when dealing with a screen presentation; simply move the display to the part of the drawing board you wish to view.

External BAM Connection Symbol.   Due to the fact that BAMs begin with a department and are thus organizationally related, parts of a flow will be contained on different BAMs. To track the flow it is thus often necessary to connect action on one BAM to that shown on one or more other BAMs. An external BAM connector is used to do this. All cross-reference information should be noted at both the exit and entry points. This information should include the name and number of the BAM, as well as the name and number of the BAM action step to which the exit or entry is connected.

If an automated drawing tool is used, the limitations of the tool and the way it connects related components should be clearly understood for larger BAMs. The ability to connect BAMs to each other, regardless of the departmental boundaries, is critical in following the flow of related activity in a company. If the automated tool cannot support this ability, external BAM connection must be tracked manually.

BAM Numbering

Each bubble is numbered consecutively, normally from the top left side of a diagram to the bottom right side. Numbering begins at the top level with 1. The second level will have a period and then a second number—1.1, for example. Again the same left-to-right numbering approach should be followed. Action 2.3 thus indicates that this is the third bubble at the next level of decomposition for the second bubble of the first level. This numbering continues until the function level has been reached. Each level will have its own number represented in each function's number. For example, function number 3.7.8.9.2 indicates that the function level has been reached at the fifth level of work breakdown for the activity. Figure 5.5 illustrates BAM numbering.

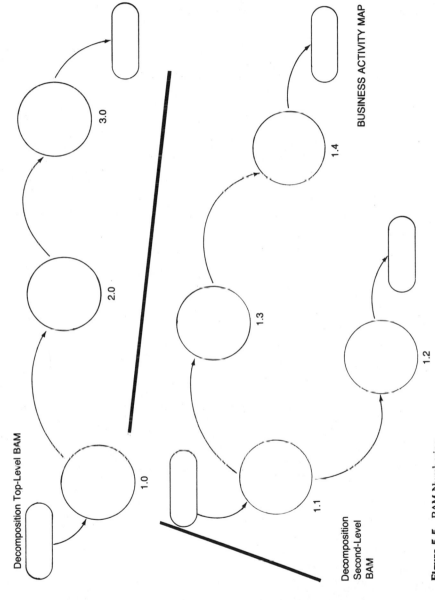

Decomposition Top-Level BAM

1.0   2.0   3.0

Decomposition Second-Level BAM

1.1   1.2   1.3   1.4

BUSINESS ACTIVITY MAP

Figure 5.5. BAM Numbering.

111

Information Associated with Each Business Function

The BAM is a document that compiles a great deal of information in an organized manner. This information includes:

1. The identification of all computer systems screens used by a business function.
2. The identification of all reports used by the business function, both computer generated and manually derived.
3. The rules and policies that apply to each business function.
4. Any external processing support, that is, credit bureaus and the like.
5. Timing or cycle information, such as peak sales, production or processing periods.
6. Descriptions on who, what, when, where, how, and why.
7. Any special activities, such as projects the applicable staff are involved in.
8. Volume information, totaled by a period and totaled per person by job classification and level.
9. Staff position descriptions and the number of people by position description and skill level.
10. Basic identification of all major tasks performed within a business function and their classification into jobs.
11. Notes of problems or weaknesses with the current manner in which the business function is being performed.

While this list is a good starting point, it should not be considered complete. Company-specific information should always be added.

Creating a BAM

The BAM is created through a series of interviews with managers and staff members. In practice, the managers will provide the information necessary for the first level or two of decomposition, and the staff will provide information for the lower level or two.

Begin the BAMing process at the department level with an interview with the manager. Each interview has two purposes. The first is to inform the personnel being interviewed about the process, why it is being done, and how it fits into the overall effort to re-engineer. Care should be taken in these interviews to begin building confidence with interviewees. Their ideas are important, and this is their opportunity to have an impact on what they will do. This may be the first time that many of the staff will be required to define what they do and consciously examine the corporate culture and their business paradigms.

The second purpose of the interview is the collection of information. All required background information must obviously be obtained in as brief a time as possible. Consideration should be given to the daily work demands of the people being interviewed. Even those who are assigned to the re-engineering effort will seldom have their workloads reduced. So their time is important. As several sessions will likely be required to build the BAM and its supporting information, concern over the use of time will help to ensure interviewee acceptance.

In the initial interview, the interviewees are asked what it is that their organizational units do. Also, they should be asked what the business unit is responsible for and what they produce. Although some analysis is usually necessary, this information is the starting point for the highest-level BAM. Next ask how these activities are related and what sequence they are performed in. This information allows the activities to be flowed. All supporting information that is applicable for the level of detail being addressed should also be obtained during the interview.

If possible, a whiteboard should be used during the interview. As the actions are identified, they can be drawn on the board. In this way the interviewee can better relate to the identification of the actions and the flow of the work. It also acquaints them with BAMs in a way that is easy for them to understand.

In questioning, the analysts continually ask how things relate and what is done next. Each branching is followed, one at a time until is it is complete. Then each of the branchings that were left as place holders on the main flow is traced. It is important that this be done in an orderly manner to avoid confusion. As the flow unfolds, ask what documents are used to do something and ask what problems they are experiencing when they perform the action. Also, when a flow is completed, ask if they have ideas for how things could be improved.

Following this approach, the interviewees usually begin to think in an organized manner about what they do. Their participation becomes increasingly more valuable as the process continues. At the same time, the analyst will learn everything necessary about their work.

Following the initial creation of BAMs for a given business area, the analysts should work together to balance the connections and cross-references. As the flow of work will continue throughout the department and the company, it is important that the analyst clearly identify all flow connectors. This is important in making certain that all flows are properly connected (what one sends, another receives; what one receives, someone sends). When the number of people who will be interviewed and the fact that they all represent different narrowly focused views of activity are considered, it is not surprising that BAMs initially have many mismatched flow connectors when put together. Balancing is thus an important quality control step in BAM creation.

When all work actions at a given level have been defined, the analyst and the interviewee will begin the decomposition process. This is accomplished by looking at each action bubble on a BAM and asking what is being done in that action. The response is noted as "bullet" points on the BAM being analyzed. When this has been done for all activity bubbles on the BAM, return to the first bubble and note the list of "things" being done. Each of these "things" becomes a bubble at the immediately lower-level BAM. The analyst and the interviewee now repeat the process of looking at each action bubble and defining its relationship to the other bubbles and the flow of activity. If additional bubbles are noted through this analysis, add them to the list of actions for the appropriate bubble on the BAM at the immediately higher level.

This process of decomposition continues until the function-level BAM has been created.

BAM Standards

Although standards must be individualized to each company, the following standards should be used as a minimum starting point.

1. Each BAM should begin with a brief description of the activity. This should describe what the BAM's action is part of and how it relates to other BAMs.

2. Each BAM should identify the originator, the date it was originated, a version number along with who made the change, who authorized the change, and the date the change was made.

3. Each BAM should be clearly cross-referenced to the organization unit that it is derived from.

4. Each BAM must be decomposed until individual business functions are identified.

5. Each action bubble should have a brief identifying name and a number that applies to the level and the order in that level's flow.

6. All flow should begin in the upper left corner of the page and move downward and to the right.

7. Each action bubble must have an initiator (document) and each must have at least one exit point (document passed).

8. Every flow leg must have an end or exit point.

9. Each action bubble must be annotated with a set of bullet items that identify the action bubbles at the immediate lower level of detail.

10. All decisions must have a "note" that briefly describes the decision.

11. Each decision diamond must have at least two exit points. Each exit must have the alternative that it represents clearly marked.

12. Each decision diamond must contain a brief statement of the decision it represents.

13. All supporting information related to an action bubble must be clearly cross-referenced to that bubble by name and number.

14. Each business function level action bubble should have information describing who, what, when, where, how, and why. All information and edit criteria associated with the bubble and all business policies and rules that are used in the bubble should also be cross-referenced.

15. Each flow connector must have all documents that are involved in the flow clearly identified on the diagram. Packets of documents can be defined by a single name. Each document in the packet must be cross-referenced to its detailed supporting information.

16. Every off-page connector must have a corresponding connector at the referenced point in the BAM.

17. Every external BAM connector must have a corresponding connector at the referenced point in the indicated BAM.

BAM quality should be controlled by a quality assurance function, which is responsible for ensuring the compliance with all BAM standards and for issuing additional standards when they become necessary. With the probability of multiple departments and divisions using this technique, it is important to adhere to a standard use and presentation format. In addition, nomenclature rules should be carefully followed. This again, is critical in producing models that are clear, noncontradictory, and usable by the entire corporation.

Relational Diagrams

Relational Diagrams are a combination of graphical representation and text that depict the flow and relationships of manually performed and automated tasks in a job. Using these diagrams, the interaction between people and computers is described in an action/reaction flow. In this flow the logic in the systems is clearly laid out in a step-by-step manner that allows it to be understood by the line manager and his or her staff and agreed upon. With Relational Diagrams, everything a person must do is specifically stated, along with everything the computer and any external supporting activity must do or provide. Because tasks are sequenced and timing requirements are noted in the task's description, everyone knows exactly what will be done and exactly when it will be done.

Relational Diagrams are used to gain a detailed understanding of how work is actually performed. It also allows change to be planned and executed with surgical precision. The Relational Diagram allows the designer of the new operation to lay out exactly how work will be performed and to define exactly what support will be required from the information services department and from external sources (such as a credit bureau or a supplier). With the help of line managers and staff, the new job's work tasks and flow can be quickly and accurately defined.

Through cross-references to supporting information, the Relational Diagram provides a direct link to policies and rules. For example, human resources decisions related to new workflow designs are supported by position descriptions, referenced in the relational diagrams, and, through references to computer system screens and reports the diagrams are also cross-referenced to detailed computer systems specifications.

Using these links, the diagrams support effective impact analysis. Because all parts of the operation are linked at this level, the indirect impacts of a change can be predicted, with related costs estimated, plotted, and dealt with in a controlled manner.

Relationship to the BAM

Relational Diagrams are designed to show how the actual work is performed. These models pick up where BAMs stop and provide a step-by-step flow of the tasks performed by each job. Each BAM business function will have identified groupings of tasks as separate jobs. Each BAM will thus have one or more Relational Diagrams associated with it to show the flow of work at this lowest level of detail. Each Relational Diagram will accordingly be cross-referenced to the BAM it supports.

Format of a Relational Diagram

A Relational Diagram is divided into three columns (see Fig. 5.6). Leftmost on the diagram is the Operational Flow column. All manually performed tasks are shown in this column, as are all physical outputs of the computer system (reports and screens). The center column is the System Activity Flow, in which all computer system tasks and their associated files are laid out. The flow of activity is shown both within each of these columns individually, and then from column to column as people

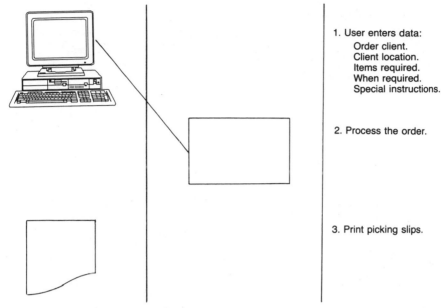

1. User enters data:
   Order client.
   Client location.
   Items required.
   When required.
   Special instructions.

2. Process the order.

3. Print picking slips.

Figure 5.6. Example relational diagram.

initiate tasks for the computer and in return are given new screens to fill in or information in some form to continue the manual processing activity.

On the far right side of the diagram is the Action Description column. This column contains a running narrative that describes all activity in text form. Each task will have an associated entry in this column. All cross-references are also placed in this column.

As a job will usually require many tasks, it is probable that several pages will be required to trace the flow. When this happens, each page is tied to its predecessors through cross-reference notation in the graphical columns and in the text column.

Each page has descriptive information that ties it to the organization it is associated with and to the business function that it helps describe. This information also contains date and version information for control purposes.

Relational Diagram Components

Like BAMs, Relational Diagrams are graphical models comprised of several standard symbols. These symbols depict action and flow as tasks are performed and computer system support is provided. The symbols used in Relational Diagrams are the:

Action symbol.

Action number symbol.

Flow initiating/termination symbol.

Decision symbol.

Computer screen symbol.

Report symbol.

Computer file symbol.

Report file symbol.

Work flow connection symbol.

Off-page connection symbol.

External diagram connector.

These symbols are a minimum set of those that might be used. It is anticipated that each company will expand this set with symbols that adhere to existing standards. New symbols that help increase either relevancy or ease of use within a company should be added through a control mechanism. The addition of this set should be made following approved standards. Each of these symbols is now discussed (see Fig. 5.7)

Action Symbol.   The action symbol is a rectangle. This symbol represents a single specific task that is performed by a person or by a computer. A brief identifying title is placed inside the rectangle.

Action Number Symbol.   Each action symbol is numbered. This number is placed inside a small circle that is positioned immediately adjacent to each symbol, except connector and computer file symbols. The number symbol is also placed immediately in front of text in the Narrative column to cross-reference the text with the symbol for the task being described. (See "Numbering Work Tasks," page 12.)

Flow Initiation/Termination Symbol.   Each Relational Diagram's flow is initiated either by a document being passed to the job being described or by an event (that is, a product is received or released from production). A flow is terminated when all tasks have been completed. Flow can also be initiated or terminated within the context of a computer system's activities; it does not need to be a manual task or event. An oval is used to represent both the initiation and termination of a job's work flow.

Decision Symbol.   A decision is represented by a diamond. This diamond is a free-standing symbol, and it is not drawn touching an action

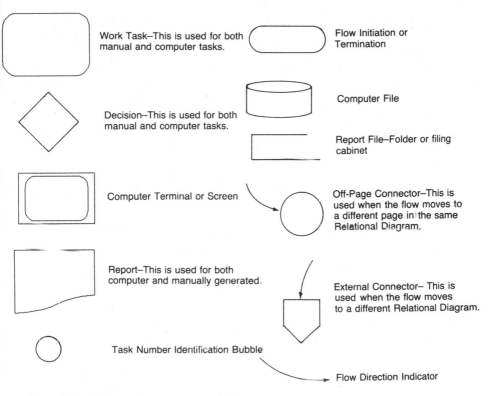

Work Task–This is used for both manual and computer tasks.

Flow Initiation or Termination

Decision–This is used for both manual and computer tasks.

Computer File

Report File–Folder or filing cabinet

Computer Terminal or Screen

Off-Page Connector–This is used when the flow moves to a different page in the same Relational Diagram.

Report–This is used for both computer and manually generated.

External Connector– This is used when the flow moves to a different Relational Diagram.

Task Number Identification Bubble

Flow Direction Indicator

Figure 5.7. Relational diagramming symbols.

symbol (as in the case of a BAM). Decisions represent conditional points in the flow where, depending on the decision, one leg or another of the flow will be followed. This thus represents an "or" condition. If more than two alternatives to the decision are possible, the decision symbol will have an exiting connector for each possible decision alternative. For clarity, each exit from the decision should be clearly labeled to show which alternative it represents. The document or product of the decision must be clearly indicated as a label to the connecting flow line.

Computer Screen Symbol.   A computer terminal symbol is used to represent a screen. This symbol is used for microcomputer, mid-range computer, and mainframe computer screens. The type of computer should, however, be noted in the narrative. The screen ID name and number used by the system it is part of should be written inside the screen symbol. Additional details should be included in the narrative. This symbol is used any time a person uses a computer terminal—regardless of the reason that the computer is being used. If a "batch" or overnight request is made through a paper request, this request would be depicted using an action symbol, indicating that no computer terminal was used for this request.

Report Symbol. This symbol is used to represent a report being generated by a person or a machine.

Computer File Symbol. A cylinder is used to represent a computer file. This symbol is always used in the System Activity Flow column (second column). The name of the file should be written inside the symbol. The system name that the file is part of should be noted in the narrative. If the file is a common file, it should be described as such in the narrative.

Report File Symbol. A small rectangle with an open side is used to represent a report file. This is the place one goes to get needed information from a manual or paper source. The name of the file should be written within the symbol. The location and responsible business unit names should also be written in the narrative. As this symbol represents a "file cabinet," it is always used in the Operational Flow column.

Work Flow Connection Symbol. The placement of each task indicates its relative position in the work flow and its relationships to other tasks. Flow is added by connecting the symbols with lines, in the order of their performance. As with BAMs, flow direction is represented by arrowheads on the flow lines. Each flow connector is labeled with an identifying name for the document or product that is carried on to the next task.

Off-Page Connection Symbol. As with BAMs, this symbol is used to represent the continuation of flow to another page in the same Relational Diagram. Again, the point of connection, page, and task number should be clearly noted. This notation should be placed on the graphical part of the diagram and in the associated text description in the Action Description column.

External Work Flow Connector Symbol. When one job's work flow interfaces with another's, an external work flow connection symbol will be used to depict the interaction. As with the use of an off-page connector, the interface must be clearly labeled. In addition to the page and task number of the interfaced Relational Diagram, the diagram's name and number must be noted. This notation should be entered in both the appropriate graphical column and in the Action Description column's narrative.

Numbering Work Tasks

Numbering begins with the initiating action as 1 and continues to follow the sequence of the flow. Each task is given a unique consecutive number. If the diagrams are drafted manually, changes or additions should be numbered as either .1 or "b," ".c," "-d," and so on. Following these schemes, additions to task 2 would be either 2.1, 2.2, 2a, 2b, or something similar. Either scheme is acceptable, as is allows the flow relationship to remain clear.

Each task number must also be used in the description of the task in the Action Description column to cross-reference the symbol to the text.

Information Associated with a Relational Diagram

At the task level, the Relational Diagram provides cross-reference correlation to a great deal of detailed descriptive information. This information is critical to the development of a sound understanding of the work being performed. This understanding is the basis for true impact analysis and for simulation modeling. All ripple associated with a change can be tracked through the relationships and cross-references shown in the Relational Diagrams. The following information is typically cross-referenced through a Relational Diagram. This information list is not all-inclusive since the information requirements will vary by company. The list is, however, a good starting point in determining the full range of information that should be cross-referenced in a given company.

1. Computer screen layout and data information.
2. Computer file data use information.
3. Report format and content information.
4. Information on who, what, when, where, why, and how
5. Business rules and policies.
6. Position descriptions and human resources information.
7. Detail on calculations and formula.
8. Detail on the design of computer systems.
9. Volume information, total and by individual.
10. Special data editing criteria.
11. Details on system and manual processing logic.

Creating a Relational Diagram

Creation of the Relational Diagram for a business function begins with a review of the function's BAM, its supporting information, and the definition of the jobs associated with that function. This provides a framework of knowledge that allows the person developing the Relational Diagram to understand the activity of which the business function is a part. It provides a context for activity.

The analyst next reviews the list of tasks associated with the job and begins associating the applicable business rules and policies, and so on, with the job and with individual tasks. With this background, the analyst will be prepared to meet with line managers and the staff who are performing the job.

Working with the line manager or staff, the analyst lays out the sequence of tasks and identifies any missing tasks. Missing tasks will be added to the task list for the job. This updated task list can serve as the basis for a position review to check on the classification of staff members.

The tasks will now be in approximately the correct order. At this point decisions and system interaction are added, and the tasks are related to one another in a flow. All required associated information is also identified as the tasks are described in the Action Description column of the diagram.

The flow will weave its way back and forth between the Organizational Flow and System Activity Flow columns as people perform tasks and the computer system reacts to data entry and inquiries. The flow may run in one column or another for several tasks before moving to the other column for a point of interaction. There are no rules as to how many tasks can be performed in any sequence of steps before an interaction point is reached.

This flow layout is continued until the tasks have been performed and the flow is either completed or exited to another business function and its associated job for continuation within the context of that function.

Once the initial Relational Diagrams are completed in the Positioning activity, these diagrams will be added to the available background information for re-engineering. Also, any change will be evaluated against these baseline diagrams. The new design will be created as a variance from this baseline. This variance can be minor or a completely new design depending on the nature and scope of the change.

Relational Diagram Standards

To promote consistency and encourage interdepartmental use, Relational Diagrams should be produced according to set standards. The following is suggested as a minimum set of standards for Relational Diagrams.

1. Always begin a flow with a clearly stated description of the conditions initiating it.

2. Each job identified in a BAM should have a fairly complete list of major tasks.

3. Only manually performed tasks and supporting logic should be entered in the Operational Flow column.

4. Only computer system tasks and logic should be entered in the System Activity Flow column.

5. Only narrative should be entered in the Action Description column.

6. All cross-references should be noted in the Action Description column.

7. All formulas for calculations should be noted in the Action Description column and cross-referenced to a detail discussion.

8. Each entry, regardless of the column, must be numbered.

9. Each screen should be labeled with the screen's system reference identification number and name.

10. Each report should be labeled with its official name and reference number. If it is created by a computer system, the system's name should be noted in the Action Description column.

11. The narrative for a task should spatially be located horizontally across from its symbol. At no time should a narrative for a task be located on a different page from the task's symbol.

12. All work flow connectors should be easy to follow.

13. Do not clutter a diagram by trying to fit too much on a single page.

These suggested standards should be customized to each company to ensure that everyone understands the presentation of information. In this way acceptance will be improved, and quality and uniformity will be promoted. If a quality assurance function is created, this person or persons will be responsible for both BAM and Relational Diagram standard development and compliance.

Summary

In this chapter the two primary techniques used in business process re-engineering, Business Activity Mapping and Relational Diagramming, were discussed. The application of these techniques will be in the next chapters as the Positioning and Re-engineering levels of the change model (Fig. 1.1) are addressed.

6
Positioning Practices

This chapter addresses the creation and use of the models that position a company to take advantage of continuing change, along with the body of interrelated information essential to the baseline of a company's operation. The following chapter, Chap. 7, will demonstrate how to use this baseline and how to re-engineer business processes.

Positioning the Business for Continuing Change

In Positioning, the first use of business models will be to provide a baseline description of the business's operation. Creating this baseline is one of most important objectives of the Positioning level of the Change Model (see Fig. 1.1, page 15). The baseline provides a starting point and a standard for measuring and defining change. These models, and the many items of information that support them, are used to analyze work flow and to identify opportunities for improvement. They are also the basis for simulation modeling and impact analysis.

In practice, the Position level of the Change Model is made up of several modeling efforts. The primary models are the Business Activity Maps and the Relational Diagrams. However, strategic plans, tactical plans, organization charts, business rules, business policies, position descriptions, production specifications, and information systems documentation are also used in Position modeling and are required to develop a comprehensive understanding the company, its culture, and its operation.

Promoting Continual Change

The models comprising the Positioning baseline work together to make continual change possible, by providing detailed information about all aspects of an operation. Through these models and supporting information, it is possible to trace all related actions as they flow through an operation. By mapping the flow and identifying the relationships, the change team will be able to determine the scope of a proposed change. This ability is augmented through reuse of the models to demonstrate the current state of a process at any time.

First, the motivations of change are identified: for example, external pressure, internal improvement, and/or technological improvements. Next, it is necessary to determine the feasibility of any proposed action. To do this, the suggested change is analyzed using the models that make up the Positioning environment and the probable impact is ascertained. Alternative change designs are then modeled.

In this process, each of the model types and its supporting information will be reviewed as the potential changes are simulated. As each simulation is tested, the change team tracks the ripple effect of the changes throughout the company. In this way, the business will understand the impact on all of its processes, its personnel, its organization, its planning, its policies, its rules, and its technology support.

This analysis is supported by the Positioning models. If the models are manual, this job is time-consuming and tedious. If an automated system is used, this analysis becomes much more straightforward and the results are expedited. Positioning models provide control to the business change process: for the first time, management will be able to evaluate proposed changes effectively.

This approach has been applied at the enterprise level, at the department level, and at the problem resolution level. While re-engineering can be performed at all these levels, experience shows that the goal of companies should be the incremental development of the baseline for the entire corporation. Only when the Positioning models are available for the entire operation can its benefits be achieved.

Sooner or later, most companies will re-engineer virtually all their operations. They will do it one piece at a time, but the aging of the operation, the change of management, and the need for operational flexibility, quality, and cost reduction will force most companies to undertake this move. The problem is that, if the traditional approaches continue, the work that is performed will provide only one-time, short-lived benefits. We believe that this is wasteful, and that, unless companies implement programs of continual improvement, they will simply lose their ability to compete effectively.

The Need for a Baseline

Michael Hammer's recommendation, "don't automate, obliterate," has become a common aphorism in re-engineering. This may be interpreted as a recommendation to totally abandon current processes. Often it is cited as the basis for omitting the development of a detailed understanding of the current business—why bother, it will all be changed! However, while everything is certainly a potential target for change, and certainly the old ways should not simply be spruced up or automated, the fundamental rationale of the business will probably not be altered by change projects. Although the way it is performed will change, the purpose of the work will remain. For example, order processing will remain a business activity, as will accounting. Even the details of the work, and the corporate policies related to it will not all change. All that is productive should be preserved.

For example, all the major activities needed to produce, market, and support a publishing business will be required for that business to operate. No amount of re-engineering will change the need to collect the material for the book, to print it, to take orders and do the accounting, and to ship it. How these activities are performed is, of course, the domain of re-engineering.

In addition, businesses are complex entities in which virtually everything is related to everything else. In this environment, the ripple of a change passing through an organization can be significant. And this ripple is bidirectional. As a change propagates outward to cause other processes to change, these changing processes themselves will have an impact on all they deal with, including the one that originally initiated the change. To avoid causing more harm than good, it is important that these relationships be clearly understood.

The failure to consider the indirect effects of change has caused re-engineering efforts to fail in the past. It is simply not enough to look at a single activity, process, or organizational unit. Limiting the view can cause a set of reactions that seriously damage the ability of surrounding departments to do their work.

Also, given the magnitude of re-engineering that most companies anticipate, it will probably not be feasible for any company to be re-engineered in one massive move. Successful re-engineering takes place incrementally over time. Each of these increments will require background information, which must be collected separately when there is no Positioning baseline. The information collected for each small effort will not, however, be sufficient to be used for modeling, and it will be difficult to keep the information up-to-date. As a result, reuse of this information is generally not possible and each effort must begin from scratch. This wastes time. In Dynamic Business Re-engineering, the models of the present operation are considered to be of paramount importance. They are formal and they are updated. Control over the versions of the available information is provided either through strict manual change standards and a change control function or through an automated versioning in a computerized support system. In looking at change, the dependencies found between departments, information, and processes require that a holistic approach be taken. This approach requires the ability to cross-reference and relate information from many areas of the company.

For these reasons, it is necessary to develop an accurate, detailed model of the operation as it exists. This provides the framework for change. It defines the infrastructure of the company in terms of its plans, its work processes, its organizational structure, its personnel, its rules, and its support systems. Without this framework, the company is a mystery to the change team. Also, assessments of new designs, analysis of increases in value, determinations of feasibility, and implementation are all forced into the realm of the hypothetical, with no standards against which to measure improvements.

What Does Modeling Show?

Few senior-level managers completely understand how the company operates at a detail level. Few companies recognize the flow of the processes used to produce products or to support the operation of the business. As discussed, the evolution of companies has arbitrarily broken processes apart along organizational lines. Department managers are encouraged to be concerned only about their own operations. Over time, the complexity of the operation and narrow incentives have caused managers to forget their company's processes. Today these processes cross organizational boundaries to interact with other processes in a complex network. But few companies have either defined or reconstructed their processes. Until recently, this information was simply not considered important; the popular approach to improvement was oriented toward departments, not process. The result is that much of the knowledge of process has been lost to companies. With this loss of process identity and the accompanying loss of a detailed understanding of the operation and its work flow, corporate management has often found that they have not had the information needed to control the operation.

Change in this environment can only be approached as a guess. No one really knows at the start what will be affected or how the change will need to be implemented. For this reason major changes have been feared. Also, because impact at a detail level cannot be determined, management has been forced to deal with problems at a gross level. The results have been catastrophic: many companies have lost their competitive edge, and hundreds of thousands of people have been laid off in mass downsizing.

It is a simple, but ineluctable, concept that what is not understood cannot effectively be improved. So the place to begin a program of continual improvement is the development of a complete, accurate comprehension of the current business' operation. In the Dynamic Business Re-engineering approach, this comprehension is obtained through the initial corporate models, which are created in the Positioning level of the Change Model. This is the first step in moving into the change paradigm. These models provide a clear understanding of the business and how it works, as well as a base for defining and analyzing change. They also allow change to be understood and support impact analysis by showing the relationships between activity and supporting information (such as business rules, policy, plans, goals, human resources, and information services). Given the analysis of the process work flows and their relationships, the ripple effect or impact of any change can be defined.

Defining Business Rules

Every aspect of a business process is governed by a set of formal and informal rules. The rules of a company are an interpretation of the company's policies at a lower level of detail; there should be a direct link between them.

The Relationship Between Policies and Business Rules

Policies are supported by business rules. This relationship is vertical in the organization, since policies are defined by rules, which are in turn defined by lower-level action or business-unit-specific subrules. Together, these rules provide governing direction on the actual manner in which tasks will be done, decisions made, and decision parameters calculated. For example, a higher-level business rule states that the company will deduct personal income tax from each employee. The procedure states that employee tax will be calculated at this point in the work flow. The detail level business rule defines the computation formula to be used in the calculation. These rules, and even the specific numbers used to apply them, are important and must be captured in the Positioning models so that they can be used in new process designs.

Business rules are applied to multiple functions and their application may cross process boundaries. To determine where a business rule is used, a matrix of all the functions and rules is needed. To do this it is necessary to normalize or standardize the naming and definition rules throughout the company. This matrix will define the relationships between the rule and its use, and show the impact of any decision that changes a rule. Also, because rules are associated with policies, it is possible to track the impact of any policy changes.

Because a rule is associated with a business function, it is carried with the function into processes modeling and then back to the new physical work flow design of the streamlined operation. This relationship allows all the business rules associated with a process to be known and questioned. It also provides guidance on how actions will be performed and a basis for audit and quality assessment.

Defining Corporate Terminology

As each change team's analysts review work flow, the terms used to describe action, reports, concepts, and anything else should be formally collected and written definitions produced. These definitions should be published as a corporate dictionary of terms. As Positioning progresses and more areas are analyzed, the number of terms will increase. In addition to collecting terms as each area is addressed, terms should be solicited from the entire company (possibly through the corporate newsletter or journal). This will expedite the creation of the dictionary and the discussion of definitions.

The purpose of this initial dictionary is to provide a catalyst for discussion. Any problems with terms should be brought to the change team's attention for resolution. Resolution should be reached through a team of managers from as many departments as possible. Once agreed on, the term and its definition will become the corporate standard.

This is not a trivial activity. Companies have a serious problem with communication. One of the primary reasons is the lack of consistency in the use of terminology. Bringing managers and staff to a common understanding of terms is a critical step forward. This is also an absolute requirement in moving into a process-based operation. As processes cross organizational boundaries, the terms used in the performance of the processes' work will change. When a multidepartment team is brought together, all must adjust their terminology to a commonly acceptable set of terms and definitions. Once this is done, communication will improve and the productivity of the team will increase.

This standardization of terms is also important to an information services operation. In the past, confusion related to terminology has plagued both the builders and users of computer systems. Because systems were developed to support specific business needs, most systems in companies today operate as independent "stand-alone" entities. Their data names and definitions are unique to each system. When new systems are added, they also add their data names and definitions. The result is that it is difficult to move data from system to system; commonly named terms can have very different definitions in various systems and the same definition can be applied to terms with different names. This has seriously impaired the ability of information services operations to provide adequate support.

Establishing common terms and definitions can help computer system support as much as it helps re-engineering.

Setting Expectations

Promoting re-engineering and controlling expectations are similar to marketing a new product. The change teams must understand the prospective customer base and create acceptable strategies. Based on this understanding and strategy, the effort must be sold to senior management, mid-level management, line management, and staff. With each group, each department, and often each person having an individual agenda, this task may require considerable tact.

Also, this is not a one-time sale. All involved must be sold on a continuing basis. With the length of re-engineering efforts, it is easy for people to lose sight of the objectives. After all, it may be several months before the first benefit is realized and then a considerable time, up to several years, before some of the people involved see a benefit. Managers and staff need a compelling reason to maintain an interest. They should be exposed to the successes that are accruing in other areas of the business, and they must be kept motivated by the promise of benefits in their own area.

One of the biggest problems in promoting re-engineering is avoiding excessive ambition. Management is oriented toward small, rapidly performed efforts. Re-engineering efforts are not small, and they are not rapidly performed. While many narrowly focused efforts have been termed re-engineering projects, the true re-engineering effort has a significant scope and objective. Setting expectations of rapid and significant return will cause a loss of management confidence and will result in unfulfilled expectations.

The alternative to a piecemeal approach of many fairly quick, but unrelated re-engineering efforts (with their associated limited one-time paybacks) is the full implementation of the change paradigm. To do this, a type of business architecture must be created, comprised of the baseline Positioning models and technical architectures. Using these models will greatly shorten the re-engineering cycle. However, creating this business architecture (which is the foundation for continuous change and quality improvement) requires time. And then, although the more narrowly process-oriented re-engineering projects will be performed fairly fast, the re-engineering will itself require still more time.

This will obviously not happen overnight. The majority of broad-scope (enterprise-wide) efforts take from two to five years depending on the size and complexity of the operation and the commitment of people and other resources to the effort. For this reason, re-engineering is a strategic commitment for most companies.

The Composition of Change Teams

Operating within the change paradigm will be based on the creation of semipermanent change teams. These teams must be empowered to do whatever is necessary to analyze and recommend changes. They will accordingly be responsible for simulating new processes, performing impact analysis, and creating implementation plans. As discussed, process flows cross organization boundaries. The teams will therefore be made up of people from every organizational unit that performs a part of a process. Team members will begin by developing an understanding of the entire process—at a detailed level.

Empowering Change Teams

The differences in many re-engineering efforts have been related to management focus and direct involvement. Managers at all levels need to emphasize the company's position on re-engineering and their role in changing the operation. This emphasis can be provided through such means as starting each meeting with a synopsis of their re-engineering effort and a brief discussion on how other re-engineering efforts are proceeding. Also, management should consider the creation of a re-engineering experience clearing group. As the efforts proceed, a great deal will be learned, both about the processes and the company. These experiences and this information should be reported to the senior managers on a frequent basis (along with status) and shared with the various re-engineering project or team leaders.

Empowerment is essential. The goal of empowerment should be to create an environment in which the change teams are empowered to look at any possible improvement to the processes they are responsible for. As the process improvements are approved, these individuals are responsible for working with the appropriate managers to implement the change. This responsibility for change is also true for the changes required by executive management as they position the company to improve its market position or to respond to a regulation. These teams are the analytical and design intermediary between department management and staff in re-engineering. Each improvement will set a new standard of operation. The department managers are responsible for meeting this standard and for monitoring performance against it and reporting problems. The change team will then be responsible for adjusting the processes in response to reported problems. In this way the change teams will be empowered to support all necessary change.

Gaining Support

Business re-engineering is the creation of a new system, a new institution, and a new way of business. Machiavelli warned that systems changes would never be popular; true to his warning, consultants who provide re-engineering support, as well as executives who initiate projects and corporate managers who support it, are frequently not well received. Too many have a vested interest in the current structure. These interests are in the form of pride, power, and comfort. Change challenges these three human desires.

The trick to successful change is to help people take pride in the new, to avoid challenging any person's personally perceived power, and to make those who are affected more comfortable with the new than they were with the old. But this is a formidable task. In the almost half-millennium since Machiavelli discussed the reluctance of people to accept change in *The Prince*, managers have not been able to overcome human nature in this respect. So the key is to make the change palatable to those who will be affected. One way to start making this shift is by changing the personal frameworks against which people measure new ideas. But not everyone will readily make this leap of faith. To begin, they must be open to new ideas—able to face problems objectively and resolutely enough to try a new way of solving serious problems.

People can also be encouraged by the elimination of fear for their jobs and by having their ideas received enthusiastically. They can become participants in, rather than victims of, re-engineering projects. They can become "change agents."

To be accepted, any new process design must address the three areas of pride, power, and comfort. All managers and staff must be involved to some degree in any re-engineering effort. Everyone's ideas on problems, unnecessary tasks, and ways to improve their jobs may also be solicited. Finally, the new operation must make everyone's job easier.

How to Build the Positioning Baseline

The first task in re-engineering is to determine the long-term re-engineering goal of the company. Is senior management interested in firefighting only, or are they interested in a long-term redirection to prepare for continual change? Both desires have their place. If a company is in serious peril, a long-term strategy for improvement will not be as important as surviving the year. This determination is thus important in setting expectations and in determining the level of quality in the re-engineering effort. For example, if short-term survival is the goal, only slight attention will be given to the eventual creation of a complete set of corporate process models. This is the realm of the "quick and dirty." Eventually, when the management team believes that the company is sound, the emphasis can shift from short-term gain to long-term improvement.

For companies that can afford a longer-term perspective, the creation of a complete set of corporate baseline models is important. The ability to reuse information and to quickly respond to market pressures is important. In this situation, responding to change is a corporate strategic direction. Re-engineering takes on an evolutionary quality as the operation is geared for continual improvement and rapid response. Each of the re-engineering projects is a part of the overall whole, and the conversion to operation in the change paradigm is a series of planned incremental re-engineering efforts.

The steps that are about to be described must be performed in both cases. The main difference is the attention paid to the retention and reuse of information. In reality, the level of detail required also changes. In the "quick and dirty" efforts, the objective is one-time immediate gain. The trade-off is the loss of the information collected. This has been the approach most followed in the past, even when long-term gain is desired. Today companies can differentiate between the two needs, and they can mix them to form a combination that makes sound business sense.

Regardless of the objective, short- or long-term gain, the effort must begin with executive backing. Senior management should form the high-level steering committee, and schedules and progress reports should be distributed to managers. All effort is directed and approved by this committee. The committee also makes certain that all managers understand that this effort is high priority and that it is fully backed at the highest levels.

Establishing the Positioning Team

Assuming executive backing, the first step is to establish the Positioning infrastructure described in Chap. 4. A chief change officer should be appointed, and a small change management group, also called the Positioning team, should be put to work. Even before the first data is gathered, the tools that will be used for Positioning and Re-engineering will need to be obtained and fine-tuned to the company's particular needs. The establishment of the Positioning team is also a good opportunity to begin to introduce the company to the change paradigm, by fully disclosing the mission of this new group.

The next step is to identify the departments to start with and the order in which to move outward. If the company has more than one location, the positioniong team should also decide how they will handle multiple locations. For example, will travel be required or can the information be collected through teleconferencing? Who will need to be involved at each location and when? How will the remote managers be trained in the new approach, and how will their cooperation be secured? Because production operations, administrative support, and information services are all separated in most companies, these considerations will need to be reviewed for each area. The overall approach will thus consist of a combination of factors from several areas of the business.

The steps discussed in this chapter apply to short- or long-term gain projects, and to local and distributed projects. The approach in setting up these efforts will, however, vary to accommodate the logistics and level of detail, but the steps will remain constant. In this way the method for re-engineering discussed in this book can be adjusted to any company and becomes the corporate change standard.

Positioning Start-Up

The environment must now be set up. This is the primary role of management at this time. The Positioning library must be created and its staff must be found. The logistics of moving information and storing it must be defined and implemented. Standards for such actions as management sign off on models and other documents must be identified and agreed upon. Standard document and report formats must be determined.

Project planning must be standardized and the approach and level of detail approved by the steering committee. The task plans must then be laid out and the information questionnaires drafted. If automated support will be used, which is strongly recommended, the systems must be purchased and the team must be trained in their use.

At this point, the team will be ready to begin.

Studying the Oroganization Chart

In all companies the organization chart is the one thing that can be counted on to be defined. Although it may be unwritten and it may be out-of-date, it will obviously exist. If it is formal, the company will have some type of hierarchical chart that defines the breakdown of the corporate structure into related lower-level units. This chart may or may not show reporting relationships and responsibilities. If it does not, these will need to be defined. Also, if the chart does not identify the location (and manager and secretary) of each business unit, this will need to be added, with phone numbers.

Through interviews with the senior officers, obtain a description of what the units do and their interrelationships. If department-level mission statements are available, they should be added to the information on the department and cross-referenced to the organization charts.

If a corporate mission statement exists, it should be collected and validated. If it does not exist, it must be created. This is performed in joint meetings with the steering committee. Final approval from the CEO is obviously required before this mission statement can be used.

Next, identify which departments are within the scope of the effort and the order in which they will be reviewed. As each department (or later, as each lower-level business unit) is addressed, the organization chart will be validated and taken to the level where jobs are identified by position description type and level. In addition, managers should define their responsibilities and authority. For example, what do they do and what do they have the authority to change? What is the extent of their budgetary responsibility? We have found that an analysis of this information can provide interesting information.

The department level mission statements are cross-referenced to the corporate mission statement to ensure that the entire company shares the same vision. Any required adjustments can be made based on an analysis of this information.

In the interviews with each manager, care must be taken to define what the department does. These "whats" become the basis for the highest-level Business Activity Map of the business unit.

Studying Work Processes

Because of their fragmentation, work processes must be identified in pieces and then reconstructed. In this part of the discussion, the identification of the components of the processes will be addressed. Their reconstruction will be discussed later in the chapter.

Process definition is based on the identification of the individual business function components. This begins with defining the activities in a department and continues with the decomposition of the activities until business functions have been defined.

Any use of a business function will be cross-referenced back to the work flow that it is performed in. This is a tie to the way work is being done. It provides a tie between the physical aspect (the operation at any point in time) and the conceptual aspect (the combination of business functions to reconstruct processes). This ability to work with process and then track change back to the department's work flow is the key to the eventual implementation of a change.

Business Activity Maps (BAMs) are used to build these models. Development of these models begins at the organization level where managers are asked to define what the department does. The response is, "We receive orders from customers by phone, by mail, and from a salesperson, all phone orders are hand-written on standard order forms. We review all orders for validity, and then we check their credit history. If OK, we enter the order into the system and then . . ." This information can be translated into graphical form by an experienced analyst (see Chap. 5).

While some of these activities are at a higher level of detail than others, this is a first cut at both content and work flow for the department's BAMs. Each activity will be sequenced and then the initiator of activity, normally a form or document, will be defined. All branching and external department connections will be identified at this point.

Working with the managers and at their discretion, staff members, each of the activity bubbles is broken into smaller component activities. Again, it is probable that differing levels of activity detail will be intermixed. This will eventually result in some activities being broken down through additional decomposition steps. As a result, some activities will reach the targeted functional level in fewer decomposition layers than others. To the purist analyst, this is a problem; it lacks stringent order and the interim documents are not at a consistent level of detail.

This lack of concern over the tuning of a level of detail should not be misunderstood. We assume that each level will be fairly close to a common degree of detail. If this is not true, the BAMs for that level will need to be changed. As the BAMs will be used from this point forward in all re-engineering to show how a change will affect the operation, they must be complete and they must be accurate. To support the needed analysis, they must also represent a fairly common degree of detail.

The result of this activity is a complete map of current activity. As this is done for additional business units, the BAMs will tie through the defined interfaces to provide a composite picture of activity for the company.

Information relating a work process bubble to a specific corporate goal should also be captured at the level of detail where it is encountered. For example, a hospital may wish to establish a remote physical therapy operation. The hospital may be setting up a scheduling capability to support this new operation. Similarly, a company may be providing phone-in order capabilities as a result of a goal to open a new customer base.

As the managers and staff are interviewed, a wide variety of information on who, what, when, where, how, and why will be defined. Any problems and unmet demands will also be noted. Any applicable business rules or policies should be captured at the level of detail at which they are discovered. This information will be cross-referenced to either the activity bubble or the business function bubble, depending on the level in the BAM heirarchy. Any computer system or other support information should also be added at this time.

Through the reports and the computer system screens, information on the necessary data for the action will be defined. Any calculations and edits that are identified should be incorporated with the other information at this time.

These models are not concerned with a detail definition of the data that is used, and they are not concerned with the flow of data. The purpose of the models is to show the relationships between activities and to identify discrete building blocks—business functions. However, information on the use of data in supporting these activities is important and is collected during the interview and document review process. All information on data use is associated with the appropriate work flow bubble.

Information services professionals must make this distinction: BAMs are not data flow diagrams. They may look similar, but they describe two very different things and they are not interchangeable. Also, the entity relationship diagram used in information systems data modeling is not applicable here. It helps define the relationships between data or other things, but it does not show flow and it does not provide a bridge to associated detailed information.

Business Functions

A *business function* is the lowest level of decomposition for a work flow. This level is reached when the analyst and the business manager stop talking about what is being performed and begin to discuss how it is being done.

Each business function will have a variety of supporting information. Among this information are the lower-level work tasks that comprise its activity. These tasks are the "how's" that are encountered immediately after the business function level is reached. The tasks needed to perform the activity of the business function must be defined at least at a medium level of detail. Once defined, the tasks are grouped into jobs. The position descriptions of the department or business unit is cross-referenced to these jobs and staff names are noted.

Relational Diagrams are used to depict the activity that takes place within each job. Thus regardless of the number of people doing the same job, it will only be necessary to create one Relational Diagram for that job. This is the lowest level of action.

All supportive information is now redefined at a low level of detail. Each task in a job will first be laid out in the sequence in which it is performed. All decisions are noted and all information files (automated and manual) are identified. The exact process is then flowed. Each computer system screen and report is noted as they are used, and each is cross-referenced to the system's documentation. Through this cross-reference, this information is tied to the information services technical architecture and the maps that define the manner in which the various systems interact (called System Interface Maps). All computer support is noted at the action by action level along with cross-references to supporting technical documentation (structure charts, data flow diagrams, and so on).

As with the activity flow development, the primary source of information is the business unit managers and staff. Other information regarding problems and observations will also be captured and associated with the point in the flow where they occur. To reduce the burden on the business unit's people, interviews should be focused and scripted. All supporting reports, forms, documents, formulas, and business rules should have been collected early in the activity flow development process. These are now associated with the tasks using them as the tasks are confirmed and their relationships defined. This approach saves considerable time by eliminating the need to reinterview when creating the Relational Diagrams.

The resulting detail on the Relational Diagrams provides a complete, very accurate picture of the business operation. Because all interfaces to other jobs and other functions are incorporated into the diagrams, it is possible to track work throughout the areas within the scope of the effort and eventually throughout the company as the models of new efforts are added to the Positioning baseline.

The result is generally a maze of interaction and flow. The complexity of the work being performed is often surprising, as is the amount of redundancy.

Corporate and Departmental Planning

Whether formally written or informally discussed and agreed on, every company is guided by strategic goals and tactical implementation plans. Regardless of the situation, the plan provides the direction for the company and through this direction the basic justification for all activity: if an action does not support the corporate strategic or tactical plan, why do it?

But to answer this question, the elements of the plans must be clearly defined. If the company does not have a formal plan, the analysts must create at least a list of strategic goals and define the objectives that tell how the goals will be achieved. These must next be tied to projects or existing actions at the tactical level. The result of these objectives and the projects initiated to pursue them is often change. But to understand the real impact, the plans must be associated with the departments and work flows they will affect.

This association is performed iteratively as the plans evolve and the planning body looks at the potential impact of any action on the business operation.

Studying Policies and Rules

As already stated, policies are supported by rules. The rules govern how activity will be performed and provide the formulas for calculations. As the rules were identified in the interview process, they were noted and associated with the appropriate BAM activity bubble. Later, as the Relational Diagrams are developed, these rules are associated with the work tasks they govern. Through the creation of the Positioning baseline models, many verbal rules will be defined and many written rules will be discarded from certain work flows and tasks. The result will be a much better picture of the way the work is really governed.

Defining rules is not unlike a quest. It is performed by reading through old memoranda, notes, and occasionally policy manuals. Procedure manuals sometimes also provide insight. However, the most reliable source is the person who is doing the work: most working rules are really interpretations of folklore that has evolved and has been handed down by word of mouth over the years. That is why at the Relational Diagram level most managers get their key staff members involved in the interviews and sign-offs.

Because rules often cross organizational boundaries, it is frequently necessary to create a consensus definition of a rule. If a consensus cannot be reached, it will be necessary to either create two (or more) rules or to raise the issue to the steering committee level for resolution.

A business policy or rule that governs how a task is performed is noted and cross-referenced to the task on the Relational Diagram.

Analyzing Human Resources

Implementation of any change ultimately will affect people. This effect must be anticipated and controlled. It is important that staff loyalty be promoted and that each person understands what his or her job entails. The formalization of this information through the Relational Diagram provides the basis for an appreciation of what each person really does. The results are often very surprising.

To cushion the blow of downsizing (or as some like to say, right sizing) it is important that everyone understand the rationale behind each action. The involvement of the staff in creating Relational Diagrams and redesigning work processes will help build this understanding. Through this involvement, they will know that decisions are not arbitrary, capricious, or punitive. Some people will refuse to accept any justification, but many more will try hard to understand. As the re-engineering progresses and people can understand the rationale behind decisions, acceptance and trust often replace fear and anger. But this will take time.

Unfortunately, people are the first point of blame for a wide range of problems affecting business. Quality is often considered to be a people problem, as are ineffective operations and inefficiency. While workers certainly have a role in creating these problems, we believe the real problems are related to corporate framework problems, a misunderstanding of process, attempts to cut costs and delay equipment and physical plant improvements, a focus on organization, and the belief that corporate mandate will fix everything. All these factors prevent people from doing a good job and rob the company from the quality and efficiency it needs to survive.

Re-engineering presents an opportunity to correct these problems. But management must accept the premise that the workers are not the enemy and work to provide the environment necessary for success and improvement. A paradigm shift in labor relations and attitude improvement can be incorporated into this process to help create a new culture. Part of this transformation is the inclusion of staff at all levels as "change agents." This goes beyond the change teams and quality circles to encompass all the company's staff. Anyone can have a good idea. All should be solicited if the company is really interested in continually improving its operation.

All knowledge and skill rest with a company's staff. This is an asset, even in companies with labor problems. The trick is to control the problem side. And in this respect staff is no different from any other corporate asset. The proper utilization of staff knowledge is one of the key elements in creating a competitive advantage.

Human resources information is associated with people and with position descriptions. This information is cross-referenced to show skill and salary information. By tying this to the Relational Diagram, information on skill and salary can be associated to business functions. See Chap. 9 for additional human resource considerations.

Bringing Information Services Under Control

Information systems documentation has long been the orphan of the industry. Many systems have poor to virtually nonexistent documentation. Also, many companies run their automated systems on a mixture of equipment, often by different vendors. These systems may be very old and may have reached a point of functional obsolescence: no improvements can be made to them; all effort must be devoted to just keeping them running. Business area managers often hear that the system cannot be made to do something when they ask for different or additional support.

In many companies the software systems do not communicate with one another to share data or to cooperatively provide support. This problem increases where the company uses multiple computers and becomes worse yet when multiple vendors' computers are used. While this is a worst case scenario, it is true of many companies.

Re-engineering requires responsive computer support. Information systems and technology in general are the great enablers of business and competitive advantage. This area may require investment. However, any investment should be very focused. It should begin with an assessment of the environment that the Positioning models depict, it should support the future direction of the business, and it should progress according to a plan.

The first step in supporting the baseline capability definition of the position models is to inventory all equipment and systems. Each piece of equipment should be listed and described. Both technical software, such as operating systems and utilities, and application software, such as order processing and inventory systems, should be inventoried. All locations of equipment should be noted and all users of each system should be listed. Along with the list of users, a brief description of each use should be noted. When performing this activity, all system documentation should be organized and evaluated; all problems should be noted. As this is being done, the users should be encouraged to discuss all problems with their support.

On the more technical side, an architecture and inventory of the current technology should be developed (see Fig. 8.2, page 200). The inventory will list all the hardware and the software that runs on it. The arrangement of this equipment, along with supporting technical information, forms the technical architecture. If computers or offices communicate with one another, the layout of the communication should be formalized as a communications architecture. This provides a framework for evaluating and changing the supporting technology. The same type of architecture should be created for all production equipment. If the production process moves from plant to plant, the architecture should include the equipment at each plant.

As it is not feasible to modify or replace all hardware and software at one time, any supporting move for a re-engineering effort will be conducted in focused increments. This means that, as support for these focused increments is shifted to the new systems, some existing systems will be replaced while others will have only certain existing support capabilities either replaced or disabled. Making these changes to the current systems will require a sound, in-depth understanding of the way the systems work, the information that may be passed between systems and the timing of the interaction.

The Relational Diagrams will clearly indicate the systems and the computer screens, reports, and support capabilities within the scope of any change. The design of the new operation will indicate if the system should be replaced or modified, and, if it will be modified, exactly what modifications are needed. Because systems sometimes share data files, any changes to data or the system, including its replacement, must be carefully analyzed to determine its impact. To do this all system interaction and all file use must be mapped. This mapping produces Systems Interface Maps (SIMs). All data must be clearly defined because each system is normally developed independently of all previous systems and each development group creates its own names for processes and data.

To control change, this area must be brought under control. If the analysis of the documentation indicates that this is not a concern, the company will be both fortunate and months ahead of other companies who undertake extensive re-engineering. However, regardless of the state of the existing documentation, the information discussed in this section must be produced. All change will need to be modeled from this information and all impact analysis performed as a comparison of change against this baseline.

Opportunities for Immediate Improvement

As interviews are conducted and BAMs are created, opportunities for immediate remedial action often become apparent. One of the requirements for immediate action is normally that the opportunities require only procedural, rule, or policy changes. Support for these changes is often minimal or unnecessary, but the results can be dramatic.

While analyzing business procedures in preparation for the selection of new Admission, Discharge, and Transfer (ADT) and Patient Accounting systems, it was discovered that admitting same-day surgery patients the evening before was routine. Doctors feared that patients would not get there early enough the next morning and would not have prepared correctly for the surgery. So patients were told to come in the night before. Many were getting there as early as 4 P.M. and required dinner as well as nursing care that was intensive if the patients were nervous about the surgery. Since the hospital would not get paid for the overnight stay, the admitting department did all the paperwork and held it until after the current system came back up after day end close around 4 A.M. Thus, there was a patient in the bed, but not in the system and therefore a major quality and potential liability problem. In addition, there was no recognition of the cost to the institution of the patient being cared for. Not even the recognition of a "patient day" for accounting purposes.

As a result of this analysis, the process was changed. These changes were able to be implemented immediately and did not require extensive modification of processes or support systems. The following changes were made.

1. Physicians were notified that past practices of not officially admitting a patient until the next morning were being changed.

2. Patients would be formally admitted, including registration on the system for billing purposes, when they arrived.

3. An arrangement was made with a nearby motel for overnight accommodations, wake-up service, and transportation to the hospital.

4. Costs of written-off charges due to this practice were recorded for future discussion with the admitting physician. Patients who wanted to come in the night before were told they would be charged.

(This example was provided by June Wesbury, a senior consultant with Morris, Tokarski, Brandon & Co.)

Opportunities like these are discovered because they are fairly obvious. Normally, they are discovered by recognizing some activity or combination of activities that seem strange or just "don't make sense." The other catalyst to finding opportunities for immediate action are well-known, nagging problems. As these problems are uncovered in the interview process, the analysts will keep them in mind when looking at the operation. As the analysts share information with all members of the change team, the entire team will be aware of the problems and they will also be aware of how work is really performed. This environment is fertile ground for ideas and ways to improve.

These ideas will first be discussed within the change team to determine feasibility. Next they will be brought to the manager of the business area or areas. Following this discussion, they will be presented to the steering committee. Because they require little effort, these changes are normally implemented quickly and with only minimal assistance from the change team.

Management Approval

Formal sign-off by the appropriate managers should be a prerequisite to continuation. Each manager who is involved should be required to for mally agree to a model or to the write-up of information that has been collected by that person and/or their staff members. This process has a tendency to motivate people to really review the information. It also forces the issue of understanding. Because people do not like to sign things they don't understand, they tend to question anything they are not certain of.

Of course, this does not solve the problem of the person who is too insecure to ask a question and verifies that the information is correct without understanding it. To some degree, because the presenter is an analyst and not a fellow manager, this problem is often negated.

The sign-off procedure should be bottom-up. The organizationally lowest-level people involved in the effort will be the ones who help with the detailed model creation and information collection. They should be the first to certify the Positioning analyses. The models and the associated information should next be presented to the immediate higher-level manager for his or her approval and signature. This process continues until a preset level of management has been reached. This approach provides the mid- and often lower-level executive managers with much more exact information about their operations than they have had. At the higher levels in a company it is impossible for managers to keep track of the details of the operation. This problem of having too broad a span to understand the details of the operation can be expected to get worse with the move to flatten organizational structure. Their review of this detailed information would thus serve no real purpose and should be omitted.

This sign-off procedure forces a type and level of communication that few companies have today. It also forces the development of a very comprehensive understanding of the business and how it works. This understanding is essential for evaluating the alternative simulation models and understanding the results of their impact analyses.

Reconstructing Work Processes

Quality and effectiveness improvement and streamlining can only be accomplished by reconstructing work processes. Problems with quality, inefficiency, and ineffective operations are symptoms of process-related problems.

However, as discussed, processes are normally fragmented and are seldom thoroughly understood (see Chap. 1). For this reason, improving a work process requires that it be reconstructed from seemingly dissimilar component parts in one or more organizational units. This involves two activities.

The first is the connection of operational work flows to provide a picture of how activities fit together. This is a starting point only. Next the business functions must be analyzed. Each business function should be reviewed and classified at a conceptual level. The question is what is this function part of? For example, is it part of what one would do to enter an order, to control inventory, or to schedule a patient for surgery in a hospital?

As each business function is analyzed and classified, it is added to a classification chart. The functions in each class are then analyzed to determine their relationships to one another. Those that do not really fit are reclassified. Those that remain are then flowed. Any holes are noted and the other classes are reviewed for functions that can be used to fill the gaps. Where none exist, you will have found an opportunity for improvement. Where you find redundancy, you have found an opportunity for immediate improvement. Using these functions and flows, the processes will be reconstructed through a trial-and-error process. The division of long flows into smaller processes is normally based on common sense.

It is important to carefully note and maintain all interfaces to other processes, not losing track of the place in which each business function is performed.

Because of the amount of effort needed to reconstruct processes and the need to keep track of all associated information, automated support is suggested. While this can certainly be done manually, it is much easier and safer when it is supported by a computer system. The Positioning and Re-engineering (PAR) system, developed by the authors, supports the data gathering and modeling of the Positioning level (see Fig. 6.1). To be fully effective, an automated system should maintain all the data required for both Positioning and Re-engineering.

Controlling Change During Modeling

Because businesses are dynamic, the collected information will probably change before it is used. It will certainly change as a result of the re-

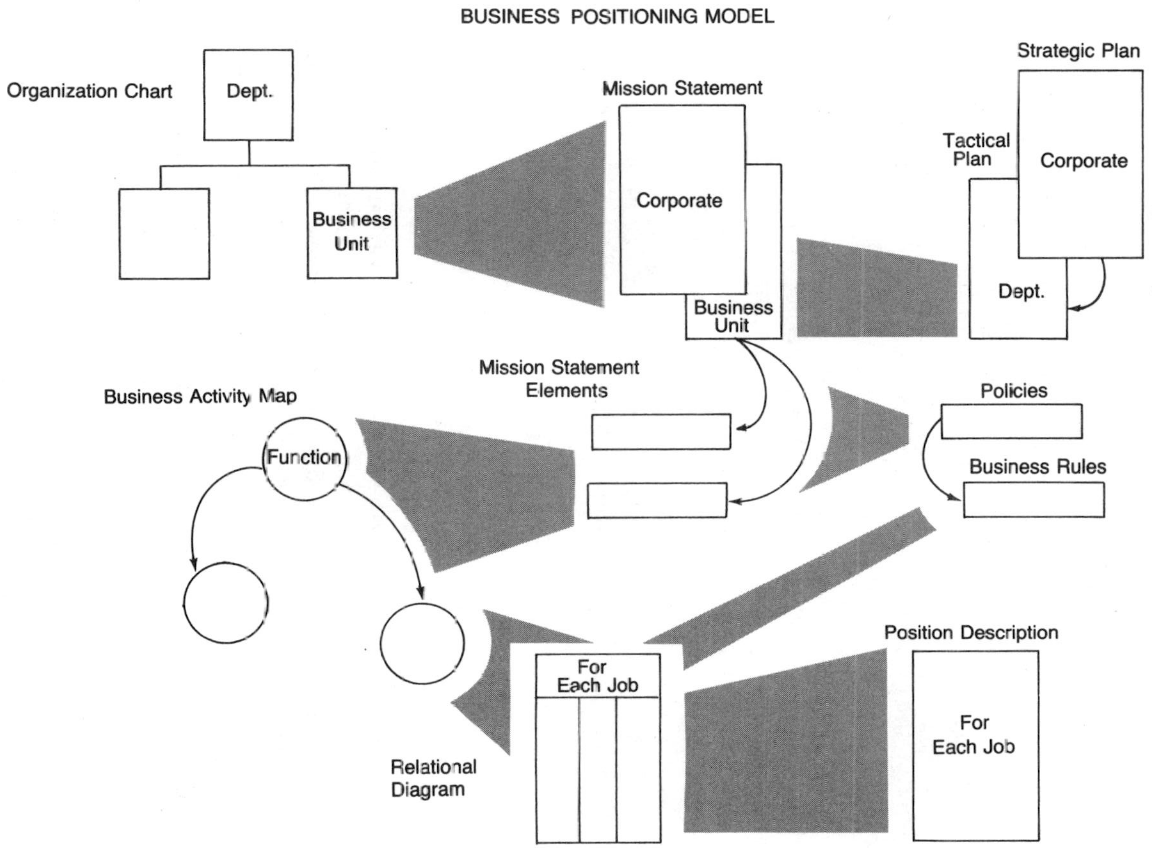

BUSINESS POSITIONING MODEL

Figure 6.1. Business Positioning model.

engineering effort, and it will change as the business continues to evolve. For this reason it is important that the information and models be continually updated. This is not a small task.

The problems begin during the development of the Positioning baseline models. As the change teams move through an operation, they will collect, organize, and store a great deal of dated information. As they move forward, the information they have collected will need to be updated. This requires the assignment of team members' time to update the documentation. As more information is collected, the amount of time spent maintaining it is increased. This reality forces a commitment decision on the part of management. Management can:

1. Abandon the information updating while the effort is in progress and gamble that the changes will not be so significant that the re-engineering is damaged.

2. Support the updating until the re-engineering in that area is completed and then discontinue the updating support.

3. Maintain the documentation both during and after the initial re-engineering of the area.

In the first two alternatives, management will be able to obtain some immediate improvement, but they will not be able to support continual change (to operate in the change paradigm). The third alternative will require the greatest commitment, but it will allow the company to support ongoing improvement.

The third option is a commitment that can consume considerable time. All maintenance of the models and information should, for greatest efficiency, be performed by the change teams. This maintenance should be supported by a computer system to help manage this information. To the extent possible, all maintenance should be performed as a by-product of daily management. Since all change should be modeled by the change teams, the daily use of the models and information will keep the Positioning baseline current.

To control change, every strategic or tactical modification to the business should be coordinated through the chief change officer (CCO). This position should be responsible for assuring that all decision information is available to the senior officers. This CCO and the Positioning team would also be responsible for the higher-level business models and would coordinate the implementation of changes into and throughout the company. In this way all managers who will potentially be involved in implementing a change will be involved in its design and implementation. Once these people have been identified, they should be brought together to discuss the change, define its scope, marshal the appropriate staff, and review the modifications that will ultimately be brought forward by the change teams.

At a lower level, the change teams, who have the responsibility of determining how a change should be implemented, will design the change alternatives, create the simulation models, determine the potential impact of the change, and plan its implementation. Again, these tasks will be greatly facilitated through the use of automated tools.

During normal operation, the business activities will change. This change is driven by improvement opportunities, technology, and unexpected needs for information or support. These changes must always be captured and any appropriate modifications made to the baseline models and supporting information. The change team will need to be involved with department management when any operational change is contemplated. Also, any time persons change something about their jobs, they should be required to notify both their immediate supervisor and the change team member who represents their business function and its jobs.

If the models will be updated using a manual process, it will be necessary to use a central Positioning library. The librarian should have complete control over enforcement of the change standards, check-out of project documents, return control, and update acceptance control. This job should not be underestimated or undervalued. This information will have been hard won through a laborious process of interviewing and analysis. It will represent a significant expense, and it is the foundation for controlling the company's ability to react to market and regulatory pressures in the future. It is also a very complex undertaking. There will be thousands of documents in a medium-sized company, all of which will be accessed and modified by a great many change teams on a continual basis.

The librarian function will also need to coordinate or take responsibility for quality assurance in the document modification process. For example, all relationships will need to be reviewed in any modification to ensure their integrity and to notify any affected teams that may not have been contacted by the change team making the modification. This is a critical step in the long-term viability of the documentation and the foundation for operation in the change paradigm.

The trade-off is in the availability of information against control over its use. Too much control means limited access, but too little risks the loss of document integrity and quality. Because no firm guidelines can given, a trial-and-error approach is suggested. In this starting approach, comprehensive use and updating standards should be developed and strictly enforced. It is always easier to relax control as needed than to destroy document integrity and then tighten control.

The Continuing Use of Positioning Models

The initial Positioning models and their associated information will provide a framework of operation for the company. As each re-engineering effort is approved and implemented, the new models and supporting information will replace the original version in the baseline model. The baseline framework is thus updated on a continuing basis.

7

Re-engineering Business Processes

After the baseline models have been created, and the appropriate information has been associated with them, the company will be ready to begin re-engineering. It is assumed that the Positioning baseline has been created, and that the change information described in Chaps. 4 (see Fig. 4.2, page 91) and 6 is available. It is also assumed that there is a change management unit (the Positioning team,) managed by a chief change officer (CCO), which keeps the company's change information, business process models and re-engineering tools up-to-date. Each re-engineering project will then consist of the selection of the opportunity, the determination of the effort's scope, the analysis of the current operation, the creation of a new design, and the implementation of the new design.

The Approach

Dynamic Business Re-engineering approaches its projects from three viewpoints. This approach addresses the three significant components of business processes, as shown in Fig. 7.1:

1. The personnel
2. The technology
3. The process itself

These three areas are addressed together, since they are highly interrelated, but they are each covered in separate chapters. Re-engineering the

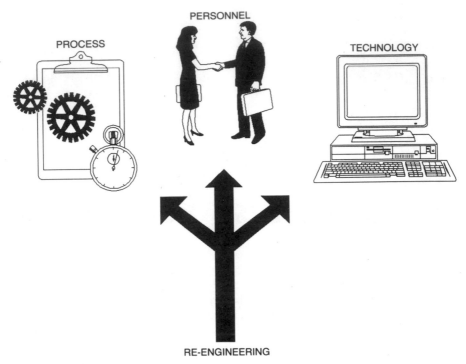

Figure 7.1. The three-pronged approach.

human resources component is described in Chap. 9. The methods used to re-engineer technology resources are examined in Chap. 8. This chapter is concerned with the third component, which is the business process.

Project Scope: The Three Levels of Change

The use of Dynamic Business Re-engineering controls change at three levels in a company. The levels at which these changes are sponsored determine the scope of the changes, as follows:

1. Enterprise-wide change is sponsored by top management.

2. Process improvement changes are proposed by the change teams.

3. Task-level changes within a job are made by the workforce, with some management coordination.

Enterprise-Wide Changes

Changes made to the entire company, or to large portions of the company, are generally made in response to external pressures, such as increased competition, decreased sales, and so on. Even when a business is operating in the change paradigm, these changes will be required, although they may seek modest results. They are initiated in the organization by top management. If enterprise-wide changes are predictable, they will be part of the corporate strategic plans. The difficulty that even small enterprise-wide changes pose is that their scope requires continued coordination from the very top of the company throughout the project at a level of detail to which the senior executives are not usually accustomed.

In Dynamic Business Re-engineering, these efforts will be performed by several change teams reporting to the chief change officer. These teams will work to determine the scope of each of the company's re-engineering efforts in terms of the business processes that may be involved in the proposed change. The boundaries of the re-engineering efforts can then be negotiated among the change teams. A single executive committee will oversee the enterprise-wide projects and address issues that cross the departmental boundaries. This committee is best formed of the chairpersons of the steering committees that control the individual re-engineering efforts.

Process Improvements

At a lower level, change may be initiated to improve a single process or a small group of related processes. This type of effort is also used to implement quality initiatives, including total quality management (see Chaps. 3 and 4). Based on observation and in response to analyses from various quality monitoring mechanisms, the Positioning team will recommend improvement efforts. These recommendations are all candidates for re-engineering projects. The efforts will thus begin as ideas and, through individual team member initiative, be modeled and analyzed. The best of these ideas will then be brought to the corporate change committee, and judged. If accepted, an idea becomes a project.

Because of their scope, these projects will impact both other processes and corporate-wide support systems, such as computer and communication support. The steering committee for the project will enroll department managers from the departments involved in the process and the departments responsible for support. Daily coordination for process-level projects can be done primarily by the chairperson of the appropriate change team. Ongoing project management will be provided by the appropriate coordination comittees. The chief change officer will review all re-engineering activity and projects for summarization and reporting to the company's executive committee.

Task-Level Change

Every worker's job is in a state of constant flux. The daily demands on most people require them to modify what they do and how they do it. These changes are obviously not considered to be company projects. They are often just subtle modifications, and they are always done on the fly. They are creative responses to the constant need to get the job done in changing circumstances.

While these changes do not require formal planning, they will benefit from coordination and control. Line managers often are aware of these changes, but because of their immateriality, no one pays much attention to them—including the person making them. They also have no impact on other people or on support. However, over time, task-level changes tend to have a significant cumulative impact. A given job may be different every few months. For this reason, it is important that the line managers and the positioning team work together to validate or modify their Business Activity Maps and Relational Diagrams on a quarterly basis. If performed frequently, this process is normally not a burden.

Incorporating Significant Factors

Re-engineering efforts may be large or small, but all the factors that determine the success of a process must be included. To satisfy this requirement, each project must have high visibility to corporate executives.

A physical therapy department re-engineering effort was conducted at a large Midwestern hospital. All processes in the department were formally identified and all work flow was mapped. All rules were also reviewed along with the department's daily problems. After study and simulation modeling, the department was redesigned. The results were significant: without a staff increase or a facility expansion, the department more than doubled its service capacity. With a backlog of patients, this was a welcome improvement. But the demand for this service was so great that a backlog of patients remained. Recognizing the potential for revenue increases, the manager of the Physical Therapy Department requested that the department be expanded to meet the patient demand. At almost the same time executive management assessed the business situation of the medical center as urgently requiring a reduction in expense, and mandated a 10-percent staff cut. Although the real reason will always remain unknown, management's decision to cut staff was applied to the Physical Therapy Department, as well as to all other departments. The cuts were made. Not only was a potential for additional revenue lost, but current revenue was discarded: fewer therapists, less revenue. The morale of the therapists, members of a profession in great demand, was also seriously impaired.

Although the attention of senior management does not guarantee success, it can provide the focus and mid-level management coordination necessary to prevent failure.

A company that sold its product in lots had a problem with the process that supported a customer's movement to larger lots. When this happened, the sales staff succeeded, but the internal operation staff failed. The process included removing the record of the current order and entering the new order. This process took more than 100 days to complete. The company was unable to isolate the delays, but they were well aware that they were losing customers. They tried to resolve this problem from an organization perspective and failed several times. Finally, they switched to the re-engineering approach discussed in this book. The entire flow of the work was traced and analyzed. It was discovered that a single change to a current order was being routed back and forth through five departments, and that over 50 people were involved. The processes were re-engineered: a new work flow was designed. Business rules were formalized; each task was reviewed for relevancy. When the streamlined design was simulated, it was found that the operation could be reduced to less than 10 days, and the number of staff could be halved. The solution included a recombination of the processes and work flow into a new organizational unit. The departments that were left with work flow gaps, resulting from relocating some of their work, were also re-engineered and improved. Finally, the new design provided previously unavailable computer support. It was found that several departments could benefit from this expansion of automated capabilities.

Both of these efforts accomplished their specific objectives. The first, however, provided less benefit than it might have, because human resource factors critical to the success of the process were decided without reference to the project. The second effort was backed by the top executives of the company who were aware of the project's goals, and removed the roadblocks to success. All three prongs of the approach: personnel, technology, and process, were included. As a result, the second project was able to realize its full potential: the problem was resolved, the staff level was reduced and several departments were streamlined. Furthermore, customer service was improved and costs were cut. Staff members, most of whom were involved to some degree in the redesign of the operation, were satisfied with their new jobs. The excess staff were not terminated, they were redeployed. Everyone won.

Clearly, two key underlying elements of success are the involvement of all levels of management, and the inclusion of all relevant variables in the project.

Re-Engineering the Operation

There are nine steps in Dynamic Business Re-engineering. These steps, shown in Fig. 7.2, provide a formal implementation of the approach just briefly discussed. They begin with the reaction to either a corporate-wide stimulus for change (market pressure, market opportunity, regulatory requirement, or technological advancement) or an idea for improvement from a Positioning team member or another staff member. The steps control activity throughout the project. While it is not possible to list the tasks that will be performed at a detailed level (there may be over a thousand in a large re-engineering effort), a general discussion of each group will provide an overview. The nine steps are:

1. Identify possible efforts.
2. Conduct initial impact analysis.
3. Select an effort and define the scope.
4. Identify business and work processes.
5. Define alternatives, simulate new work processes and work flows.
6. Define the potential impact of each alternative.
7. Select the best alternative.
8. Implement the selected alternative.
9. Update the Positioning baseline models and information.

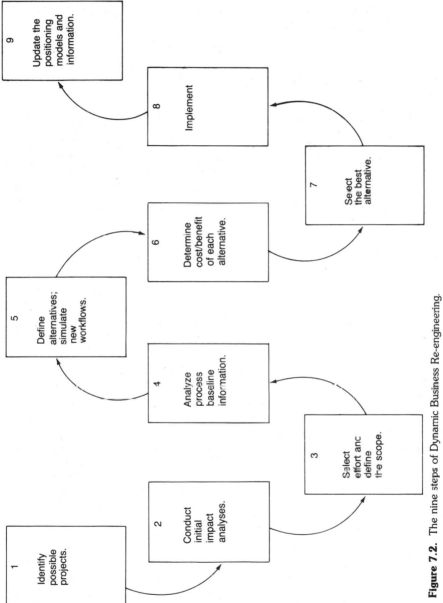

Figure 7.2. The nine steps of Dynamic Business Re-engineering.

1
Identify possible projects.

2
Conduct initial impact analyses.

3
Select effort and define the scope.

4
Analyze process baseline information.

5
Define alternatives; simulate new workflows.

6
Determine cost/benefit of each alternative.

7
Select the best alternative.

8
Implement

9
Update the positioning models and information.

The activities required to successfully perform each of these nine steps are discussed next. Each of these steps is presented in four parts: (1) a general overview of the step's purpose, (2) the deliverables of the step, (3) the major tasks that must be performed, and (4) a discussion on some of the more significant factors that must be considered in the step.

Step 1: Identifying Possible Projects

An important responsibility of the chief change officer's Positioning team is to identify the potential improvements that will inspire re-engineering projects. The team will find possibilities in their own continuing analysis and in the ideas of the company's whole workforce. The chief change officer will also receive suggestions for change from the company's top executives and even the board of directors.

Identifying Enterprise-Wide Projects

The most obvious source of enterprise-wide change is specific instructions from top management. The other source of these changes is corporate strategic plans. All successful companies try to understand their marketplace and plan their activities. Also, all companies have business plans, formal and informal. However, these plans are often not fully shared with the lower levels of the company. The Positioning team must, of course, have access to this information. To do this, senior management must be willing to discuss strategies and tactics with them. This communication is necessary even in companies that have very detailed written plans. Certainly the existing plans provide a good starting point, but that is usually not enough. There are two reasons: (1) business plans become out-of-date quickly, and (2) the many unrecorded thoughts that went into creating the plans are important. This is expected, and can and should be dealt with. To obtain current information first-hand, it is recommended that the Positioning team take the initiative and interview the company's senior officers and any others who contributed to the plans.

Identifying Process-Level Projects

Projects undertaken to improve individual processes are the preferred approach for re-engineering. They are also identified by the Positioning team, but they originate in the team's own work and in the suggestions of others in the company. To develop rudimentary ideas into project possibilities, the team will again obtain its information from interviews, but in this case from department-level management downward. Departmental plans and interviews help to evaluate every business function in a process being studied for improvement. If any function does not support a goal of the company, why is it being done? Ultimately, work should always relate to a business goal. This step allows a cursory review of these relationships.

Determining Project Objectives

The objectives of a change project will vary depending on the driving force behind the change. They cannot therefore be taken for granted. These objectives become the success factors of the project. They are the basis for evaluating the outcome of the effort and must be well defined.

In determining objectives, a framework is first created from the conditions behind the effort. This framework may be somewhat unfocused initially, since it will be necessary to perform this step at a fairly broad level. The analysis will, however, become increasingly focused as additional information is obtained and the scope of the project is refined. If the goal is to provide an efficient new process design, it will be necessary to understand the work at a detailed level. If, however, the project is intended to solve a critical problem as quickly as possible, a briefer analysis must suffice. In this case, the result will be a short-term, interim benefit with the assumption of a more permanent re-engineering project to follow.

It is important that this initial definition of project objectives include the perspective of all appropriate managers. Differences often exist even among the people who agreed to the change request. By obtaining information and background from all these people, a consensus can be pieced together. This is helpful in defining objectives and in managing expectations. If some managers believe the effort to be minor and it is not, their perception will ultimately be that other people dragged their feet. This is the first point of interaction between the senior managers' steering committee and the re-engineering project team. Expectations and objectives must be honestly set.

To do this, the discussion should focus on the consensus definition of the effort and on the initial estimations of impact, in terms of the areas of the business that will need to be modified. It should not yet focus on estimated time or dollars. For example,

> Our initial analysis indicates that 27 work flows and 57 people will be affected, 5 computer systems will probably need to be changed, one fairly significantly. We will need to extend communication support into two areas in the building at 1401 Main. Marketing will need to look at the way they handle widget commissions, and we will need to determine how the sales territories will be affected.

To this information it will, of course, be necessary to add the "bottom line" cost and the elapsed time the project will require, but, again, this should not be emphasized until later in the effort (Step 6) when the cost-benefit report is presented. This presentation relates an impression of magnitude much more effectively than either dollars or time alone.

With this approach, expectation and perception can be effectively managed. However, the initial analysis cannot be significantly off. While some inaccuracy is acceptable at this point, care needs to be taken to be certain that the impact will be close. This is one reason why the accuracy of the baseline information is so important. A quick analysis of the baseline models (Business Activity Maps, Relational Diagrams, System Inventories, System Interface Maps, and other tools) and information discussed in the previous chapter will provide a confident initial estimate—if the data is up-to-date.

Once the objectives and initial estimates of impact have been approved, the effort will be ready to continue.

Focusing Change: A Specific Set of Requirements

Every re-engineering effort will be performed to meet one or more objectives. Each objective is tied to a business goal and thereby derives its reason for existence. At a lower level, each objective will have a specific set of requirements: what must be done to meet this objective? Some common requirements are:

1. Support a part of the company's business plan.
2. Cut the time it takes to do something.
3. See if something can be done with fewer people.
4. Handle a new line of business.
5. Solve a problem.
6. Improve the standards, and thus the quality of a process.
7. Improve a service—such as customer support.

These requirements determine the focus of a re-engineering effort. They are also the factors that success will be judged against. As the incremental re-engineering of a company will continue only as long as the individual efforts succeed, it is important that expectations and evaluation criteria be formally set.

A re-engineering effort was progressing according to plan. As it was a strategic direction, it had not been formally cost-justified; senior management agreed that it was what was needed to shore up a slipping market share. A new Chief Financial Officer was brought into the company about a year and half into the re-engineering effort. Being conscientious, he believed that a formal cost-benefit analysis should be performed. Because the effort was strategic in nature, it could not readily be justified as a whole (strategic directions cannot readily be cost-justified; the tactical projects that make them up can and should be cost justified). However, failing to understand the difference between the two, he hired a large consulting firm to conduct a study. Because the study was based on an erroneous assumption, a compromise approach was required. The firm came up with one: the cost justification study went forward based on immediate dollar benefits, with no value given to improved market position and customer satisfaction. Even so, the projected savings was several million dollars per year. However, the CFO, along with representatives of the consulting firm, succeeded in convincing the senior officers of the company that the cost of the project could be better spent, and that the re-engineering effort should be stopped.

Clearly, the new CFO had not accepted either the concept or the objectives of re-engineering. Equally clear is the fact that this financial executive, while trying to use the available budget in the best ways, dealt a death blow to the company's quality and efficiency improvement ambitions. As a result, staff morale throughout the company was almost destroyed, and the company did not improve. Today the company has resorted to mass layoffs to cut costs. The procedural problem underlying this example is that the re-engineering effort did not set a specific goal and the effort's requirements for success were poorly defined. The senior officers did not know exactly what to expect of the effort and, when presented with convincing financial arguments, had no counterproposals. There was no agreed-on way to evaluate either the status of the project or the acceptability of its projected results. In good times this is a serious proposition. In hard times it presents an unacceptable risk.

Justifying projects while trying to determine requirements is a very serious and continuing problem. It is difficult to accurately estimate the eventual benefits of an environment in which change can be handled quickly and easily. It is also difficult to anticipate the results of any re-engineering effort in the early steps of the project. At this uncertain stage, managing by success criteria is a better approach.

The exception to the justification problem is the need to comply with government regulation. This will probably not be cost-justifiable. It will clearly have a single benefit; the regulating authority will allow the company to continue doing business. It may be best to view strategic actions similarly: their principal benefit is that the company will continue to be competitive.

At the process level, smaller re-engineering efforts will be geared to provide a specific benefit. They will eliminate a certain redundancy or bottleneck in the operation. They may even be commissioned to eliminate unneeded activity. These efforts are interestingly seldom cost-justified. However, their goals and requirements are also normally fairly well defined.

Finally, each goal and requirement must be approved by management at all levels. For the reasons already cited, it is important that managers concur that the effort is worth doing, and that they agree on its needs and success factors. This agreement should also be in writing. Each lower-level manager should sign a project definition document which will then be presented to the next level. In this way the project should retain sufficient support to carry it through all its steps, even if management changes.

Where to Start: Selecting the First Project

The initial effort must succeed for its own sake, and also provide confidence in re-engineering so that a program of continual change can be instituted. It is thus important that the initial effort fit a fairly specific set of criteria:

1. The effort must be easily defined; the scope in particular must be easily determined.

2. The scope must be broad enough to provide a significant benefit, but narrow enough to control easily.

3. The initial effort must not be too difficult. The method and techniques will be new to the group; they will need experience before tackling complex projects.

4. Those involved must want the project to succeed and must commit the resources necessary for success.

5. The effort must have senior management's participation.

It is also recommended that projects that are especially subject to political and personal agendas be avoided as first efforts. Since the initial effort must prove the worth of the approach, the first project must be particularly compelling. Indeed it is best if it satisfies all the criteria and also has very attractive business objectives. These objectives should seem ambitious for any change project, but this is especially important for an initial one.

Subsequent Efforts

Following the initial effort, each re-engineering effort must be individually justifiable. Many efforts will be proposed from the sources mentioned. The number of possible efforts can be expected to exceed the capabilities of the company, after the first project succeeds. This will probably create a backlog of re-engineering work.

A two-part review is suggested. The first part will be performed in this re-engineering step. This is a review of the proposed change and its potential benefits. Those that are chosen as possible projects will be passed to the impact analysis review: re-engineering Step 2. Project selection, in Step 3, will be based on this analysis.

Approving the Re-engineering Project

Change is constant, but only some of it is good. To control and justify change, a formal procedure is followed by most businesses, which incorporates executive approval of all change projects. This procedure applies to all corporate projects; there are no special considerations for re-engineering. The concept of having a chief change officer arises from Positioning, and provides that the CCO will be responsible for the formal change procedure and the presentation of all change projects to the executive committee.

Step 1 Deliverables

The deliverable product of this step is the initial evaluation of requests for re-engineering projects, along with a definition of the objectives of each effort, a definition of the effort's specific requirements and an assessment of the nature of the effort—process improvement or broad-based re-engineering.

Step 2: Conducting Initial Impact Analysis

A simple impact analysis should be performed to provide an initial understanding of re-engineering requests. Each project that is passed forward from the first step will be considered to have a sound potential for acceptance. These projects will be reviewed against the current Positioning baseline models, to determine their potential impact on the operation and the company.

With the exception of determining the initial organizational boundaries, the approaches taken in determining the initial impact of widespread re-engineering efforts and of process-level improvements are virtually identical. The differences are in the amount of information that must be dealt with, not in the steps followed.

The review and subsequent analysis should first identify the departments that will probably be involved in the effort. This will set the initial boundaries of the impact analysis. Next the requirements of the change should be used to determine the processes that will be involved. This is done by reviewing all conceptual processes in each affected department and determining which of them will be affected by the requirements of the re-engineering effort. At this time a quick evaluation will be made on how the process will be affected. Based on this identification, the department list can be revised; all departments that perform a part of the affected process will be included in the effort.

A high-level review of plans, policies, and procedures for the affected departments will provide an initial idea of the full extent of the effort. The greater the proportion of policies and procedures that are affected, the deeper the impact of the project. In addition, the possible effect on information systems support, on communication systems, and on production capabilities should be assessed.

This review is cursory and is performed to provide a list of potentially affected areas. No detailed investigations are performed at this time. An initial rough cost-benefit estimate can also be provided. These estimates will indicate order of magnitude: the cost will be very small, moderate, etc. The same will be true for the benefits of the project. This information will allow management to understand the magnitude of the effort and gauge the differential between cost and benefit. Since the primary goal of this step is simply to weed out the efforts with high probable impact and cost, but low benefit, it is suggested that as little time as possible be spent in this step. The rough gauge will normally allow management to make this decision. For those efforts that pass the tests imposed in this step, the definitive scoping and impact analysis will continue in Step 3.

Step 2 Deliverables

An analysis of the probable impact of the project on each affected department's organization and work flow, on all business operation processes, on business rules, on information services support and on staff will be performed. This analysis is used in this step to determine which projects merit further consideration—and which are to be passed to the next step. The deliverables of this step are a list of re-engineering projects that appear to be worthwhile and their associated impact analyses.

Step 3: Selecting the Effort and Defining the Scope

Obviously, the selection of re-engineering projects will be based on benefit. However, benefit cannot always be determined in the traditional ways—cost recovery or avoidance and sales potential. For example, it is difficult to quantify the dollar benefit associated with intangibles such as improving interaction with customers who have problems or who need other forms of assistance. The initial impact analysis will, however, help quantify some factors. It will certainly help to make cost estimates more accurate. It will also provide the initial forecast as to what streamlining will do for the process and the work. But these estimates will be soft. The early steps of re-engineering analyze problems, but the benefits lie in the solutions to the problems. Until the solutions can be designed, any estimate of benefit is speculative.

With use, as more direct experience is gained in re-engineering, a company can develop some opinions concerning the average benefits of streamlining and improving process quality. These should provide credible forecasts for future proposed projects. For example, if an analysis of those operations that have already been re-engineered show a range of between 15 and 27 percent in cost savings, the case could be made that this range will also apply to similar work in most other operations in the company.

As with all other processes in the company, the procedure used to evaluate and select efforts for re-engineering must itself be constantly monitored and adjusted. Goals can be set for re-engineering, just as they are for other corporate activities. Performance should then be monitored against these standards. Constantly apply what is being learned to this process, and the process of re-engineering itself will improve the speed and quality of the effort, including the accuracy of its estimates.

Setting the Initial Scope of the Effort

The scope of a re-engineering project is the boundary of the process that is to be re-engineered. It is not, as is stressed throughout this book, defined by organizational boundaries. Furthermore, it must encompass an entire process: while all the work flow of a process may not be changed, it must all be included within the scope of the project. Setting scope is therefore not straightforward, but it is very important. Setting the scope of the initial re-engineering project is particularly critical because, if the first effort fails, it is not likely that a second effort will be undertaken. The first scope-setting activity is also difficult because the company will not yet have learned to visualize processes in preference to organizational structure.

Once the initial project has succeeded, the first victory in a winning record will be in place. All subsequent efforts can be selected based on impact, but the scope of these efforts must be carefully controlled. Complexity grows dramatically with scope. The scope should thus be broad enough to provide a real benefit, but focused enough to be controlled. As with all other activities, it is much easier to succeed at small, simple efforts than at large complex ones. In Dynamic Business Re-engineering, the initial scope estimated in Steps 1 and 2, is much broader than the project will actually address. In this step it will be refined and approved in its final form.

Amoeba Scope

Amoeba scope is a process-oriented technique used in defining the scope of a re-engineering effort, which recognizes the haphazard evolution of businesses (see Chap. 2). Processes in most companies today are arbitrarily divided along organizational lines, although the process work flow is known to cross them frequently. If all the departments that are potentially involved in a process are laid out in a hierarchical organization chart, the boundary of the process work flow would wind its way from unit to unit to form an irregularly shaped object, as is shown in Fig. 7.3. The form of this object will change continually as the flow of the process changes. The picture is one of an irregularly shaped object that is slowly, but continually, changing: an amoeba.

When defining the scope of an effort, the starting place is a review of the current Business Activity Maps and Relational Diagrams. The flow should be followed and all the applicable business functions identified. The boundaries of the flow will then be used to define the scope of the project: the process (or processes) to be re-engineered. A quick look at the

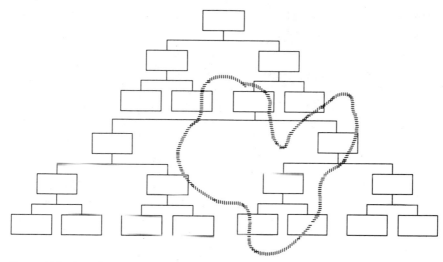

Figure 7.3. Amoeba scope.

interrelationships between processes will identify any other processes that may be affected, and will provide a fairly comprehensive impact assessment. Because the models and change information maintained by the Positioning team relate the company's functions to its organizational structure, the departments that will be involved in the project can be quickly identified. These lists of departments and processes define the true scope of the re-engineering project.

This approach circumvents the need to distinguish between a symptom of a problem and a problem before setting the scope. However, if the company requires that a department, instead of a process, be re-engineered, some additional work will be required. First, it will be necessary to identify all the business functions the department performs. From these, it will be possible to identify all the business processes in which the department is involved. At this point, all other departments involved in all these processes must also be added to the scope of the project. So starting with a single department is much more complex than concentrating on an individual process.

Step 3 Deliverables

The most important deliverable produced by this step will be a list of projects, selected from those that emerged from Step 2, that are to be scheduled and executed. Also deliverable at the end of this step is the formally defined scope of each of these projects.

Step 4: Analyzing Business and Work Process Baseline Information

The engineering work of re-engineering begins with this step. The first three steps were involved with selecting the area of the business to be re-engineered and defining the scope of the project. At this point, the boundaries of the project will have been set and approved by the appropriate managers. This step's activities include defining the models, developing the information that will be required, and analyzing the work flow.

While this may seem to be straightforward, it is not. It involves scrutinizing the policies, business rules, costs, values added, revenues, work flows, conceptual business process models, business functions, organizational structure, organizational unit missions, job definitions, production processes, and computer systems that are related to the process. Tracing interrelationships, quantifying the models, and determining information requirements can be very complex activities themselves. Finally, the familiarization process, which acquaints the change team with this information, requires an intensive effort. This step is, however, the foundation for the re-engineering; the effort is worthwhile.

All the previous information and analysis will be further refined in this step. This refinement will provide more accurate estimates of the potential costs and benefits of the project. Also, the probable effect of the project on work flows will be described, and the appropriate managers will be brought together to form the project's coordination team. Most important, Step 4 develops fully detailed models of current processes to enable new processes to be designed. This is the step that must identify all the problems with the way things are being done.

Problem Correction Projects

For those efforts performed to resolve a specific problem or a group of related problems, the Step 4 analysis is started by determining where the problem is encountered and its symptoms. If a quality assurance program is in place, its statistics will provide a good source of symptom data. This provides a point to begin investigation. Go backward in the work flow from this point. Identify the work flow that produces the symptom: a scrap heap, a rejected product, a billing problem, and the like. Next, identify all the activity, relationships, and problem areas in this work flow from the Business Activity Maps.

Identify all the business functions in the work flows that seem to have problem areas. Next, examine the Relational Diagrams to learn the details of the work being done. Note the technology used to support the business functions. Track the problem back through the work flow to determine if the problem is caused by something done in another process. Also check for obvious weaknesses, which may cause or contribute to problems. These flaws are often procedural in nature, and typically involve tenuous communications, incorrectly defined work, fragile work procedures, lack of precision in task definition, poor technical support, and the like. Once the causes of the problems (and potential problems) are isolated, the analyst will notify all business units involved in the process. The analysis should be verified by managers and key workers.

Process Improvement Projects

Unlike projects that are started to resolve problems or streamline a department, process improvement projects normally begin with the identification of an opportunity. As a result, the initial Step 4 activity is to analyze all process interfaces. This will again be used to identify potentially affected departments. Next, trace the business functions comprising the process back to related work flows and departments.

The Positioning models that help to identify opportunities (see Step 1) also provide the basis for analyzing the detailed current work related to a process improvement. Scope is not difficult: by their nature, the scope of these projects will usually be narrowly focused, which keeps complexity to a minimum. But, as with all re-engineering efforts, the project will include modification of departmental work flow, business functions and Relational Diagrams (jobs) conceptual processes, and often computer and production systems. These projects thus require the detailed analysis of all relevant Relational Diagrams and supporting information.

The challenge in process improvement projects is to assure that all affected activity and support are considered. Because they begin with an opportunity, which is more a solution than a problem, these projects may be subject to insufficient analysis.

Enterprise-Wide Re-engineering

Enterprise-wide re-engineering projects differ in that they will involve several, and possibly all, of the departments in a company. Corporate plans, which have defined most of the major corporate-wide projects, should also have identified all departments that will be affected and determined their roles in the project. If not, this step must begin with these definitions. Also, the list of departments in the corporate plans may need to be verified against the Positioning models.

Once the departments who will be involved have been verified, the same procedures used for organizationally oriented projects will be followed. The primary variation is in the need to continue the coordination of the project at the highest management level, and to combine the work of several change teams. This difference is one of management span and not methodology.

Adding Quantification to the Models

Adding the numerical data related to the business process is done, in most cases, by annotating the Business Activity Maps. The BAM has two basic components, the arrows representing the flow of effort and the circles representing actions. The flows can have only one quantity associated with them: delay time. The activities of a process can have many metrics. The metrics most commonly associated with process activities are as follows.

Cost. Cost is, of course, the most important measurement of any business component. The difficulty lies in determining the cost for an individual activity. Since these numbers are used for comparisons, consistency is more important than absolute accura Simplicity in estimation is suggested. For cost, the material cost per unit processed is usually available. Staff costs can be salary per labor hour multiplied by the hours spent per unit in the activity. These figures may require some estimation or measurement. The overheads can usually be safely allocated based on labor costs, unless the physical plant is one of the key change factors being studied by the re-engineering project.

Input. The amount of material or number of parts required as input to the activity is shown in two ways: (1) per unit of output and (2) current requirements per unit of time. These figures are generally not complicated, except in cases such as chemical processes, for which the batch is the unit, and many types of chemicals may go through the same processes, generating long lists of input and output requirements.

Output. The amount of output of the activity, in terms of parts, batches, or transactions per unit time. This is the payoff of the activity, but it is usually uncomplicated and not difficult to obtain.

Time.   This is the rate at which input is turned into output, usually expressed as just the output rate.

Head Count.   This, the actual number of staff performing work functions in a given activity, is related to labor costs, and it is required for change project analysis and design.

Quality.   Quality is measured in many ways, and the ones most meaningful to the process should be used in the models. For example, rejection rate is usually included in both production and operational activities. Also, any customer-driven quality measurements should be included, if they are available. When an activity seems to have no quality measurements at all, quality must be defined and meaningful measurements developed.

Value Added.   Value added can be entered as either the actual value added to the product or service, or as the time spent by the staff that actually adds value.

There are also metrics related to service agreements (agreements between the Information Services department and the business operation departments that define what support can be expected), service activities, and the contribution to staff morale made by the process design. These measurements are discussed in Chaps. 8 and 9.

Metrics can be added to models developed on paper by simple annotation or added to automated models as notes. The PAR system (see Chap. 6) has special detailed information entry screens to accommodate quantification for processes and activities. Once entered, the best test of the validity of this numerical data is to add like measurements across the entire process and the entire business, and then check the results against the business's overall head count, cost, and production data.

Step 4 Deliverables

The detailed analysis of the processes to be re-engineered is the basic deliverable of this step. The work carries the Positioning models and data to additional levels of detail and refines the data so that problem areas and interrelationships are visible in full detail. In addition, the analysis of Relational Diagrams and other supporting information will provide a full understanding of the operations and how they really function. This knowledge is the basis for the creation of new designs.

Step 5: Defining New Process Alternatives: Simulating New Work Flows

Alterative new processes are designed in this step. This includes solving the problems discovered in the last step and producing new models and new work flows. Also, when appropriate, this step may produce new organizational structure designs. Furthermore, there is a high probability that computer support will be affected as the jobs that support each work flow are modified to reflect the required changes in the work (see Chap. 8).

This step uses the information collected and studied in the previous step. At this point, the work flow and process analyses performed in the last step are used to create new process and operation designs, and to simulate the new operation. These simulations will be used to determine which new design is best. There will be as many simulation scenarios as there are significantly different ways to do the required work. The final selection of a design alternative will be made in the next step.

Who Is Involved?

Regardless of the reason for the re-engineering effort, the change teams will be responsible for creating the new process designs. For enterprise-wide projects, the chief change officer will usually direct the change teams in this effort. For department-oriented efforts, the department manager will serve as the chairperson of the project's coordination committee (other departments affected by the project will, of course, be represented). In this case, the department manager will direct the creation of the new process designs. Problem resolution re-engineering efforts and process improvement efforts will be managed by either a coordination committee or an individual department manager, depending on the scope of the effort and the departmental boundaries that are crossed.

In all cases, department managers must participate in the redesign work. Their participation includes advising and providing insight into the corporate culture and subtleties of their operations. They also provide a manager's "sanity check" throughout this critical step. The pragmatic viewpoint of line management is useful in making the best compromise between the current work flow and the new designs from the change teams. By direct participation, the department managers have the ability to encourage creativity while monitoring the project's impact.

Creating New Designs

Several actions are required to create a new process design. First, the overall process work flow is revised to achieve the specific goals set for the project in its early steps. The changes are then reflected in the work flow of each of the affected departments. Next, the new work flow in each department is optimized for performance. In this way the process is redesigned and the work flow in the departments is optimized. These together form the new operation. Organizational changes are considered when the departmental work flows are optimized.

The key to creating this new design is the modularity of the business function. Because each business function has been cross-referenced to all its supporting information, it can be treated as a building block. As the work moves from business function to business function in the process models, its path is shown on the Business Activity Maps. The details of the business functions can be changed using the Relational Diagrams associated with each function. Change can either be applied over the entire process by moving business functions or addressed within the confines of each individual business function. Of course, due to relationships, business functions that are not deliberately changed by this procedure may be affected, and may require some adjustment.

New work is actually designed at the task level. The details of work are reflected in the Relational Diagrams; new work is constructed by drawing new Relational Diagrams. The Business Activity Maps provide a higher-level flow and structure to the design, but they do not tell how the work will be done, or how the process will be improved. Each redesigned Relational Diagram depicts all the old and new tasks and their sequence. All computer support actions are also included. By a simple comparison of old to new, the exact changes can be defined.

The change teams usually do not have difficulty producing several new design alternatives for each process. The work is done using trial and error, but the choices are generally clear if the current operation and its supporting systems have been well researched. This is, of course, the creative part of re-engineering. The project's participants can apply everything that they have learned about what works and what does not work in their company, and in their business—all of their ideas, experience, and training.

Creating New Business Activity Maps

The Business Activity Maps for a single new process design alternative will be redrawn several times. First, the functions making up the processes that are the subject of the project will be restructured. Then the impacts on other processes will be assessed, and their BAMs will be redrafted. Third, the details of the work will be changed using the Relational Diagrams, usually requiring the BAMs to be redesigned again. It is essential to use some form of automated systems to help draw these charts.

If starting this step is difficult, it is suggested that a "strawman" design be created by combining all similar job descriptions in the process into one function for each and then connecting them to form a work flow. For example, all secretarial work would be put into a function called "secretarial work." This organization of the work will almost never be desirable, nor will it be, in fact, even feasible, but it should provide the absolutely least-cost alternative. The change team is usually stimulated by this alternative to come up with other, feasible ones.

Care should be exercised when restructuring BAMs to assure that all the exits, or outputs, from each business function, and all the entry points, or inputs, are accounted for in the new designs. If any are eliminated, it must be done intentionally. If any are forgotten, there is a very high probability that the model will fail. The damage may be significant if this occurs after the project has ended and the model is being used for production.

Creating New Relational Diagrams

In creating the new design, each decision and task in the Relational Diagrams must be reviewed for continued relevancy and for improvement. Staffing requirements must be analyzed in terms of both work volume and skills. The new designs will address the project's goals and will reflect both the support available from technology (see Chap. 8) and the corporate culture. Creativity is the principal ingredient in re-engineering. However, change should never be made unless it significantly improves something. So, if the new design is not superior to the old, the change team should try another approach.

As each business function's jobs are modified, it is important to change all supporting information. For example, the context of the change must be documented for future change teams and efforts. This is the who, what, when, where, how, and why information. It explains why the change was made and describes all design decisions.

It is important that the workers who perform the business functions that are being changed participate in the job redesign activities. Their participation provides detailed insight. For example, they may raise performance and ergonomic concerns that are often missed. As one of the goals of re-engineering is to make their jobs easier and smoother, this insight is invaluable. The other advantage of this approach is that it promotes acceptance and commitment. The workers feel that they are important to the company and that the promises made regarding re-engineering are being kept. This, of course, promotes trust in management. Finally, this participation provides an education in work flow analysis and in the use of Relational Diagrams that allows the workers to help control and document the small continual changes to their jobs.

Changing Department Work Flows

Because the changes are made in the business functions, they will change the departmental work flows. Therefore, changes to the process's work flow and relationships will also be applied to the appropriate places in departmental work flows. In addition, work flows must be reconstructed around any gaps that are created by moving functions. Finally, as the departments work flows are being redesigned, additional streamlining and quality considerations are applied to the new designs. In this way all the process changes are applied to the departments, and the departments' operation is improved.

Determining Changes in Organizational Structure

If corporate reorganizations are required to maximize the benefit of a new process design, this requirement must be demonstrated to executive management. This can be done by testing alternative organizational structures using the process designs: determining if it makes better sense to put similar business functions in one department or leave them separated. When the fragmentation is high, significant gain can be made in combining the activity into a new business unit.

Major reorganizations will create personnel issues. They should be avoided if the same benefits can be obtained by moving responsibilities and work flow, without changes in organization structure. When reorganization is appropriate, it may imply relocation, staff reassignment, training, and new computer and communication support services. For these changes, a new organizational model will be required. Mission statements must also be modified or created. The business functions, like building blocks, are not likely to be severely affected by a reorganization. Their supporting "where" information will, of course, require updating.

The Simulation Model: Validating, Simulating, and Analyzing Results

The models used by re-engineering are not simulation models in the sense that, if a data item is changed, the results of the change are not computed automatically. However, the models can be used to simulate changes in work flow. The models for each alternative design can be evaluated manually, using the quantitative data associated with the work (see "Adding Quantification to the Models," page 170). Since these evaluations are made to detailed function and process designs, they can be very accurate; they are certainly sufficient to judge the alternative re-engineered designs.

The initial analysis of the new design models will be performed at the process level using process BAMs and then at the business function level using the Relational Diagrams. This first test is a high-level validation: does the change make good business sense, and does it successfully address the goal of the project? If these tests are passed, consideration must be given to the implications on other processes. Questions regarding the assimilation of other business functions from interfaced processes must be reviewed. If the answer is positive and some business functions will be either assimilated or deleted, the impact on these processes must be considered. The gaps left in the other processes are called "black holes." Unaddressed, they will cause serious disruption in the process and in the work flow of the department that performed them.

Additional tests that should be made are:

1. Looking for redundancies in activities and processes.
2. Looking for bottlenecks in work and task flows.
3. Identifying ineffective operations.
4. Identifying inefficient operations.
5. Llooking for the reasonable resolution to causes of business and support problems.
6. Making certain that all interfaces are sound, that the outputs of each match to the inputs of their destinations.
7. Making certain that technology capabilities are being fully used.

Obtaining Approval

If multiple design scenarios have been created, they should first be reviewed with the personnel in joint review/design sessions. The purpose of these sessions is to select the best alternative from the worker's perspective. An interesting aspect of these sessions is that, based on their review, it is not uncommon for the designs to be broken apart and parts from several recombined to form a new design that is acceptable by all the affected workers. This new design must obviously be tested before the next action is taken.

The recommended solution and the other simulation alternatives will be presented to the coordination committee. These managers will be "walked through" each Business Activity Map and each Relational Diagram to show how the process and their operation would be affected. This presentation should be in a joint session, in which all the members of the coordination committee are present. Consensus in determining the solution to be implemented is important. Only in extreme cases should the need for a consensus decision be overridden by senior management.

Step 5 Deliverables

The product of this step is one or more detailed re-engineered simulation scenarios that depict new process designs. The designs include:

The redesign of the appropriate processes.

The redesign of business functions, job tasks, job work flows, and position descriptions.

The design of computer and communication systems enhancements.

The redesign of the departmental operations work flow.

The creation of new rules and policies.

These models and associated information are created for each process and department affected by the effort.

Step 6: Evaluating the Potential Costs and Benefits of Each Alternative

At this point one or more simulation scenarios of the new operation will have been developed. Standard measurements will have been applied to these designs to help determine the amount of improvement that can be expected. The costs and benefits must now be specifically defined before a recommendation can be made.

For the most part, this step uses standard cost-benefit analysis. As most managers have lived with these studies throughout their careers, we will assume that our readers have considerable knowledge in performing and using these studies. The following discussion will be directed to considerations that we wish to highlight or that apply to re-engineering efforts as a wrinkle to the normal approach to a cost-benefit analysis.

Identifying the Impact of a Change

The first step in defining both costs and benefits is to confirm the change team's understanding of the results of the re-engineering project. In this way a re-engineering effort differs from a traditional cost-benefit analysis. This confirmation is essentially a review of the process and interface lists to make certain that all extensions were considered. Following this check, the exact extent of the change that the new design will cause will be defined. Each change to a work flow, business function, process, job, or support must be considered. The degree and nature of each change can be used to cost that change. As the individual changes are aggregated into higher-level arrangements, the costs will be grouped for presentation.

The results of moving from impact estimates to a full understanding of the impact throughout the operation will significantly improve the ability to determine both cost and benefit. Since management will have a comprehensive list of all the parts of the business that will be affected and a firm understanding of how each part will be affected, the estimates will become more detailed and more accurate.

Determining the Probable Costs of the New Design

There are two different types of costs associated with each new design: the one-time cost of implementation and the continuing cost of operation afterward. The cost of implementation, added to the cost of the re-engineering project, will be the investment that the company will make in the new process.

Clearly, the usual cost items will be considered. Some of the more important of these line items are:

Salary and personnel overhead costs.

Computer and communications systems costs.

Production equipment upgrade or replacement.

Ancillary technology upgrades (such as hand held inventory scanners, etc.).

Physical plant changes.

Moving personnel and their equipment.

Some business disruption or possibly interruption costs may also be determined and added to the direct cost considerations. Personnel retraining and possibly outplacement costs should also be considered.

As each aspect of the business that changes will produce a cost, each process, business function, job, and department work flow that is changed must be reviewed to determine what it will take to implement the change. This analysis is the basis for cost determination and provides the starting point for implementation planning.

This costing will consider both direct and indirect costs. Because of the level of detailed information available, more complete and detailed estimates can be derived.

In looking at costs, it is suggested that a new item be added to the corporate chart of accounts: knowledge. By considering the replacement costs and the ongoing investment in education by both the personnel being educated and those who will help them, the costs can be calculated.

The average by skill level and job classification can be applied in re-engineering, especially if staff cuts seem appropriate. Experience has shown that staff cuts may be only temporary; staff levels creep back to some degree. The savings sought through cutting the workforce should be tempered by these considerations.

Defining the Expected Benefits

Some benefits of re-engineering will be tangible, while others will not. As with all other cost-benefit comparisons, benefits can be divided into two categories: those that can be quantified (such as cutting waste or time) and those that cannot. However, the intangible benefits may provide the greatest long-term impact. For example, improving customer support will have tangible and intangible parts to its benefit. Similarly, improving product reliability will certainly save return and repair costs. It will also promote good will and improve customer loyalty. While it is fairly easy to place a dollar value on the first benefit, saving return and repair costs, it is not easy to assign a dollar value to improved good will and loyalty.

In re-engineering, it is often the intangibles that provide the most compelling reason to implement a new operational design. In the long run, improving product reliability and customer satisfaction will provide the highest benefit—certainly more than cutting a cost or eliminating positions from the payroll. For "bottom line" managers this concept is often foreign; accepting this category of benefit as valid represents a paradigm shift for these individuals.

Where possible, a dollar value should and will be assigned to each benefit. All assumptions and decisions associated with this assignment of value should accompany the report to allow the decision body to understand the basis for the calculations. The typical categories of dollar assignment relate to time savings (streamlining), waste/rework savings, raw material savings (quality improvement), and staff reduction. The nebulous improvement savings are related to gaining market share and include increased sales volume and opening new markets.

Cost-Benefit Analysis

The goals of the re-engineering activity should be factors in the cost-benefit analysis. When viewed on an individual basis, clearly return on investment becomes an overwhelming factor. Here the objective is to save money. However, if an individual effort is simply part of a long-range corporate strategy, this expenditure becomes an investment. Because each effort must have a compelling reason for its implementation, the distinctive and separate goals driving each effort must be clearly defined and recognized.

This study, along with the underlying reasons, will produce a recommendation as to which alternative should be chosen. An equally valid recommendation to cancel the project will also be considered based on this study. If the project is simply too costly or would require too great an upheaval, a recommendation to either put it on hold or scrap it might be presented by the project's management.

In normal circumstances, one alternative will be clearly superior to the others. To save time and money, the designs that are considered to be inferior will have been dropped in the last step while still partially completed. All scenarios that are moved along to this next step should be valid. If this "weeding out" is truly accomplished, the cost-benefit analysis becomes a very important tool.

In performing the cost-benefit analysis we urge all participants to be open-minded and fair. We have all seen these studies skewed to prove a position or back a specific solution. In many companies this is ingrained into the corporate culture: they are proving what they want. To avoid this situation, we suggest that the intangible but direct benefit of supporting a long-range direction of the company be given additional weight in evaluating the study results.

Step 6 Deliverables

The primary deliverable of this step is a detailed analysis of the costs and the benefits that are associated with the implementation and use of each simulation scenario of the new operation. The final product of this step is a recommendation of the scenario that should be implemented.

Step 7: Selecting the Best Alternative

The approach used to select the best alternative will vary in each company. The differences will be related primarily to corporate culture: each company will view empowerment differently and each will have a different comfort level with allowing staff to make decisions. Regardless of the selection approach, the selection of the best alternative will be related to benefit and cost. This is the greatest benefit, with the least impact and the smallest cost. In addition to these criteria we suggest that a third gauge be used. This is the ability of the new design to make jobs easier and free staff from mindless drudgery. This will improve loyalty and morale, and ultimately improve performance.

Who Will Select the Alternative?

The selectors will change with respect to the scope of the effort. With large high-impact efforts, the senior officers will need to have final selection authority. For department-oriented change, process improvement, and often problem-resolution-oriented change, the coordination committee or an appropriate vice president or two, will probably make the final decision. For efforts that remain internal to a department, the department manager will select the alternative.

This selection by higher authority does not contradict the move to empower staff with decision-making authority. It rather recognizes that authority limits will be set and works within this context to define responsibility and authority. This differentiation is critical to successful empowerment.

The Selection Procedure

The selection review opens the alterative designs to comment and potentially to change. This potential is controlled by the chief change officer, but it is not eliminated. As managers become more familiar with the design, they may notice opportunities for improvement that have been overlooked. The managers will provide a different perspective and offer a different background from which to evaluate the design. Also, because the business is dynamic, newly surfaced considerations may change the needs and thus the design.

Any required changes must be put through all appropriate impact analysis and design method steps. As these changes may significantly modify the costs and/or the benefits, the cost-benefit analysis must be reviewed and updated where necessary. For this reason, it is important that all changes be clearly defined and all reasons for the changes be captured in detail. In these situations it is important to limit the amount of rework and the number of times the design is presented to the approving body.

Again, this review should be open and old paradigms should be cast aside. The objective is to select the best alternative, not just to change things. However, if the person or group with selection authority fails to find any alternative acceptable, the change team will be required to begin again, this time with additional insight into what may be acceptable.

Although it can and has been done, appealing a selection decision to a "higher authority" is at best risky and should be considered only when the benefits are clearly compelling.

Step 7 Deliverables

The selection of a design simulation for implementation is the deliverable of this step. All affected managers and staff should be notified of the selection as soon as possible. This notification should include project timetables and information on any changes to the original version of the design selected. This notification begins the next step: implementation.

Step 8: Implementing the Alternative Selected

The detailed implementation cost-benefit review, performed in the last step, will have determined what major activities will be required to implement the alternative. This activity definition is the starting point of the implementation plan.

Creating the Migration Plan

Implementing the changes associated with re-engineering projects is a complicated undertaking. The most difficult part of any change, people, has been addressed throughout the previous seven steps in this methodology, and is covered in detail in Chap. 9. The personnel who will be affected will have worked with the change teams, first to define the old operation and then to redesign it. They will understand the techniques and tools that have been used, and they will have contributed at a personal level. This participation usually breaks down barriers to successful implementation and provides a type of ongoing training. At this point the staff will have a firm understanding of the new operation and how they fit into it.

This participation must now continue as the steps that will be used to create the new operation are defined and work is assigned.

To be effective, the migration plan must address every action necessary to build the new operational environment and then move from the present operation to the new one. This plan is thus very detailed. Each person's role must be considered and all tasks assigned. Control over this process is provided through the coordination of personnel, task relationships, and technology acquisition.

The aspects of the new design's implementation that will take the longest to complete should be addressed first. These are normally technology-related, involving the acquisition of computer hardware and software and the acquisition of communications and production equipment. The next area is generally construction. This includes the purchase or construction of new facilities and the modification of existing structures.

These parts of the plan often represent the critical path. For this reason, they should be laid out before any other activities are considered. In doing this, lay out the main actions and then determine what is needed to make each task happen or to support each task. This allows the planner to back into the schedule from the critical path. Once this is completed, the tasks needed to complete the development of the new operation and its support must be carefully defined and then laid out on the project's time line. Next, all tasks necessary to move the current operation to the new one must be defined at a very detailed level and then laid out on the schedule. Together these tasks represent the entire migration.

Once this is completed, the plan should be simulated and tested for omissions and errors. This test will be accomplished through a manual walkthrough of each step. All project participants should be included in this review. As the goal of the walkthrough is to identify problems and resolve them in a modeling environment, all participants should try as hard as possible to find fault and prove the plan will fail. When a point is reached where no problems can be identified, the plan will have passed the test.

The Components of the Migration Plan

The objective of the migration plan is to provide continuous control over the implementation of the new business operation. To do this, the plan addresses:

1. Physical plant changes.
2. Moving into new spaces.
3. Implementing a new organizational structure.
4. Implementing the new work flow.
5. Changing production in the plant.
6. Changing computer support (see Chap. 8).
7. Testing the new process.
8. Establishing contingency operation plans.

9. Training personnel.

10. Changing policies and business rules.

11. Changing and/or creating position descriptions.

The "Belt and Suspenders" Approach to Implementation

Experience has proven that no amount of planning or testing will prevent problems. And somehow they always occur at the most inopportune time.

Recognizing this fact, the plan must have built-in contingency arrangements. It must allow the implementation to occur in parallel with the current operation wherever possible, and it must provide the ability to back out of the commitment to the new and continue operating with the old. This is a methodical approach that commits slowly and always allows the company to stop and revert. Final commitment is a big step, and the full switch-over to the new operation will occur only when the new operation is fully functional and stable.

In this way problems are either avoided or dealt with easily. It has been claimed that an implementation that is performed without problems indicates overanalysis, excessive planning, and unnecessarily detailed preparation. This is a shoot-from-the-hip attitude, and it is wrong. Sloppy implementation and migration can cause serious problems for the company. By addressing these problems with well thought-out contingency plans, risk is reduced and costs are minimized. This is the "belt and suspenders" approach.

Implementing the New Operation

The first rule of implementation is to *be ready to change the implementation plan*. The business is dynamic and the operational realities change constantly. Also, even the best plans can contain errors. It is far better to admit to an error and correct it than to try to make it work.

This potential need to change the implementation plan is one of the reasons for contingency plans. For example, a situation arose in a company engaged in a re-engineering project due to an external problem, requiring that virtually all nonoperational effort be stopped. Because a "belt and suspenders" approach was being followed, project implementation activity was quickly halted, and the new process was temporarily suspended. This action did not impede the company's ability to respond to the emergency. Later, the implementation was restarted and completed successfully.

Because of the need to accommodate change, the implementation plan must be a flexible document. Certainly, changing it must be controlled and only absolutely necessary modifications may be allowed, but the ability to adjust to the realities of the implementation is a critical success factor in any re-engineering effort.

To offset the moving target that is produced by the constant changes in an operation, all noncritical change must be temporarily frozen during implementation. This stabilizes the environment and reduces the complexity of implementation.

Changing the physical environment—moving personnel, setting up a local area network (LAN), setting up the desks and phones, changing the organization charts, and producing new telephone books—is a frustrating, but vital activity. The frustration derives from the fact that this area of activity is susceptible to missed schedules. One reason is the interdependencies that must be dealt with. For example, the LAN cannot be installed until the cable is run. The cable cannot be run until the area is evacuated, the evacuation must wait for people in two other departments to move, and all effort is dependent on the electrician's contract negotiations which are in progress. And on and on. The coordination and tracking of progress are important in this part of the implementation; so is the flexibility to adjust the schedule.

In implementing the new operation, the company will have an opportunity to improve relations with its personnel. Human relations concerns should be used as a basis in announcing promotions, reassignments, training, and testing skills. The manner in which these announcements are given can make a big difference in the quest to gain worker confidence and support (see Chap. 9).

All support services and related equipment use must be coordinated during the implementation. In most circumstances, this support will have been integrated into the new design. This integration produces a very different approach to business, but also means that, when the support capabilities are unavailable, work will slow significantly. For this reason, these support services must be in place and fully tested prior to any staff move to the new operation.

When implementing the new work flow in a department, make certain that all interfaces to other processes and other departments are functioning properly. Re-engineering efforts typically involve two or more departments. As activity is concentrated into fewer work steps and processes, care must be taken to make certain that any gaps created in the work flow of other departments are addressed.

When implementing a new operation, the change team should be reoriented from project participation to continued improvement of the conceptual process. This shift is necessary to begin the program of continual improvement.

Finally, the best advice that can be offered is that, when implementing the new process, test it, test it, test it, and be ready to suspend it. Be cautious. Frustration over delay is bad, but can be dealt with, especially if the overall implementation is smooth. Operational interruption and quality degradation caused by implementation problems is not, however, acceptable for any reason.

Step 8 Deliverables

The primary deliverable of Step 8 is the migration plan. The other planning components that support the migration plan, such as new policies, organization charts, and job descriptions, are also prepared in this step.

The most important deliverable of this step, however, is the new business process itself. At this point, the new process will be operating and giving increased benefits to the company.

Step 9: Updating the Positioning Models and Information

Following the implementation of the new operation, all supporting documentation must be added to the baseline information for the departments and conceptual processes that are in place. This addition is an updating of some documents and a replacement of others. If automated support, such as the PAR system described in Chap. 6, is used, the new working models are used to replace the older version. In this way the project's models and information become the new baseline and nothing is wasted or lost.

This reuse ability relieves a great deal of the administrative overhead associated with keeping the models and information current. Because the change is designed using the models and the new models replace the old ones, maintenance is largely a by-product of the re-engineering activity.

In manually supported re-engineering environments, sufficient staff must be assigned to this task. Also, strict use and modification standards will be needed along with the authority to enforce them. In these environments it will be impossible to maintain all the information and track anything but major changes to all their interfaces. For this reason only certain models and documents will be able to be supported. Although it will vary somewhat by company, a decision on which models and information must be maintained will need to be made. Associated detailed information will become out-of-date, and, although it will provide background, it will not provide the level of support that is available through an automated system or group of systems. Changes to this detailed information may be made by the change team members. However, this will vary and, unless mandated, the changes will probably be made to personal sets of this information kept by change team members.

In both the manual and automated document storage environments, a library function will need to be created and staffed. This is the central location of all models and information. It is responsible for information and model currency, quality, and availability. All use should be coordinated with or by the librarian. This is especially true in the manual environment where all rule enforcement is in the hands of the library staff. By contrast, much of this police work will be taken care of by the automated system. This frees the library staff to provide more meaningful document location, methodology, and study assistance.

Whether the environment is manually or computer-supported, the updating of models and information will be performed by the change teams. No one else should be given update authority. Thus while any person who is authorized to do so may look at the models and information, only a specific group will be able to change them. This authority should also be confined to the conceptual processes and organizational entities for which each person on the change team is responsible. In this way quality and content can be managed and consistency is controlled.

Although historically the library function in companies has been a fairly low-level job, when operating within the change paradigm, it becomes an extremely important activity. The past handling of documentation in most companies points only too clearly to the need to document almost everything about the company. This activity is the basis for the positioning baseline creation and then for ongoing re-engineering. If any re-engineering is undertaken, the company will be collecting this background information. Since it is a by-product of the re-engineering efforts, it is virtually free. Certainly it costs a lot to create it, but, given that this cost is born by a specific project, the documentation that is normally discarded after years of storage can be an asset and can be used over again to save time and money. This reuse is made possible by the orderly storage and update of this documentation as it is used in change projects to improve the operation.

Step 9 Deliverables

The deliverables of this step are the updated Positioning models and updated positioning data, both ready for the next re-engineering project.

Project's End

Can there be anything in business as satisfying as a completed re-engineering project? Not only is the business more competitive, which is the most uplifting improvement that a company can experience, but there is a heightened sense of participation and achievement. There is also the sense that everyone knows more about the company and what makes it work. The sense is real.

8

Re-engineering Information Technology Resources

Although it is often linked to information services, business process re-engineering is not a computer activity. Some of the techniques of re-engineering come from information systems development experience, but others come from industrial engineering and many other management sciences. Furthermore, many successful projects that have been labeled "re-engineering" have been information systems projects for which some re-engineering of business processes was done. The links between re-engineering and information services may lead to the erroneous conclusion that Positioning and Re-engineering are information technology methodologies. In fact, both are business activities. Separating the concepts of computing and re-engineering is important because re-engineering projects should be the responsibility of line and executive managers, not the company's information services department.

Although re-engineering is not an information technology topic, business itself increasingly depends on computers. The use of information technology to improve the operation will therefore almost always be considered in re-engineering projects. In practice, as re-engineering projects examine business processes, they often discover new and improved uses for information and technology. Re-engineering is also uniquely able to relate the use of technology directly to the business process. Additionally, information technology should be used to help the re-engineering effort itself. Dynamic Business Re-engineering places a heavy emphasis on the use of automated models of business processes and automated Positioning support tools.

Information technology is a factor in all the layers of the change model (see Fig. 1.1, page 15). Current technology support and the design of an overall corporate information architecture are found in the Positioning level. The requirements for new computer systems are developed in the

Re-engineering level. The purchase and programming of new systems is in the Infrastructure level, and the implementation and use of these is in the Operations level. This chapter discusses how re-engineering can use technology, and how it allows business to find the best uses for technology.

Using Technology in Business

Using technology in business has been an increasing problem since the turn of the century. The first tabulating machines furnished limited capability, so it was easy to understand how to use them. With increasing speed, the capability and complexity of information technology has expanded. In the 1960s and 1970s, professional computer technicians were responsible for computing. As a result, much of the complexity of programming could be ignored by managers. Since 1980 the use of information technology has spread to every part of the business. Although it has become more complex, much of it has been placed in the hands of end-users. Today, not only is business perplexed over how to operate the technology, but also how and when to apply it.

The use of large mainframe computers has generally been carefully cost-justified, although much of the justification was based on anticipated payroll cost avoidance that was never realized. Since mainframes were a centralized cost of large magnitude, the investments related to each of them could be scrutinized at the highest levels. The use of desktop and smaller-scale technologies has changed these practices. There is generally little study done when a personal computer is bought. Recognizing this situation, management has become much more concerned about the effectiveness of information technology. Is it a productive tool or a toy?

Concerned that there was little known about the productivity of technology, the topic of return on computing investment was studied at MIT. This work has failed to find a direct correlation between information technology investment and increased profit. From these results, it would seem that the worst fears of skeptical managers are justified. Investing in information technology may be a waste of money. The MIT studies, however, did reveal a relationship between information technology spending patterns and bottom-line improvement. Companies that spend their technology budgets in certain definite ways seem to get more out of the investment. The most important factor in the use of personal computers and office technology in general was revealed to be the ratio between the purchase of technology systems and the provision of expert help to users. These findings add empirical support to the widely held opinion that it is not how much technology that is used, but how well it is applied.

This is the basis for the benefit that re-engineering can provide to the use of technology in business. Re-engineering shows opportunities to use technology, and then lays the foundation by applying the right technology for the work.

A Key to Business Process Efficiency

Technology is one of the most important keys to improving efficiency. But how does technology support a business process? How is efficiency enhanced? How can the gains be realized in cost savings? The first concern is the fundamental contribution that technology can make. The ways that technology can support a business process follow.

Increasing Speed.   Technology can be used to do something more quickly than a person. It can also decrease the elapsed time on the critical path of a process.

Storage and Retrieval.   Technology can store information and retrieve it later very quickly and with as much organizational and search capability as may be required, but at increasing costs for increasing capabilities. Technology can perform functions in this area that cannot reasonably be expected of any workforce.

Communicating.   Technology can move data and information from one point in a process to another virtually instantly, and in a variety of forms.

Controlling Process Tasks and Improving Quality.   Technology can directly control tasks in a business process. In general, this increases the quality of the output, since human error is eliminated and automated equipment can provide much finer measurement and manufacturing control than a person. For these reasons, controlling processes by the use of information is very well established in industry. Technology can also be applied to improve office and knowledge worker processes involving the complex transactions and decisions. The automated management of office control information has been pioneered by Terry Winograd and Action Technologies, in the development of The Coordinator (discussed in Chap. 5).

Monitoring.   Technology can compare what is being done to a set of standards, either as the process is being performed or afterward. Immediate problems that are reported can then be corrected and the monitoring function can test them again. Statistics regarding quality, perform-ance, use of supplies, and process results can be produced by the monitor-ing function, and these can themselves be monitored.

Supporting Decision Making.   The data required to make business decisions can be gathered and used at a decision point in the process to help the staff make better decisions or, in some cases, to make them automatically. The data can be presented in convenient forms, such as graphics, to make the decisions process easier.

Fabricating, Manufacturing, and Delivering Services.   Technology can perform actual work steps in processes of all sorts.

Supporting Process Work Functions.   Technology can assist workers in ways other than increasing speed and improving quality. Often it is possible for automation to reduce the price of an effort by the automation simply being less expensive than labor.

These categories are few, but many products and combinations of products give business the capabilities that each of the categories cover. But how is management to select the most appropriate technology? Furthermore, how can proposed investments in technology be evaluated? The answers lie in evaluating the contribution that each technology product can make. Technology by itself has no value. Without knowing the precise effect that technology will have on the work being done, it is not possible to assess its potential value. Although most businesses use information technology, few have sufficient information about their work processes to analyze its impact. It is for this reason that the benefit of technology to business remains undetermined today.

When conducting re-engineering, the information required to analyze technology contributions will be at hand. Therefore, the ideal time to specify new technology support is during re-engineering projects.

Current Technology Support

The most recent advances in information technology, and the current methods of its use in business, should be known within the company and available to the re-engineering project's team members. The situation today is a mixture of rapid technological advancement, and an intense struggle in businesses to resolve the problems that this advancement has produced. Information technology in business is therefore in a state of flux. This is primarily due to the rapid evolution of both the personal computer and data communications. The current character of technology in business originates as much from rapid technological change as from technology.

There are currently two major types of information technology: (1) mainframes (including mid-range or minicomputers) and (2) desktop or personal computers (PCs). They have been at odds with each other since the PC was introduced into business about 1981. This has often led to confusion regarding the relationship between the two. Clearly, the PC and the mainframe have different roles in business. Mainframes and minicomputers are best used to process what is known as transactions, which are groups of data that are each related to a type of business transaction. Mainframes have been best suited to this work because of their limited ability to serve a single work location. They have, over the last 20 years, developed special software to process transactions that guarantee the integrity of each transaction. PCs started by giving all their attention to a single worker, and the software that has been written for them addresses text, graphics, and numbers, but not transactions. Another restriction associated with the use of PCs for transaction processing is that they are not very secure; so transaction processing is somewhat risky. The result is that much of the transaction-processing work performed by business has remained on mainframes. This continued use has resulted in the total demand for mainframes remaining high in most large companies. In the near future, several factors will bring these two technologies closer together, so that the information technology of the next generation will be neither PC nor mainframe, but will have the characteristics of both.

PCs and mainframes (along with the enterprise-wide communications networks associated with mainframes) should be made to merge, to provide a complete and consistent information service. But today most corporations allow departments to manage their own office technology, or not to manage it, if they prefer. The current status of office technology is that it is not very efficiently used; staff spend far too much unproductive time trying to make it work properly, and the data in it can be lost to the company very easily. Mainframes and minicomputers are generally under firmer control, but years of poor success with developing systems for this technology have left management with a certain distrust of it.

In response to a growing demand for improved information delivery and communication support, the mainframe manager's interest is now shifting from application support to providing data communications network support for both mainframe and office technology. Networks are the common ground of both and will provide short-term benefits as well as the path to the future of corporate information technology.

The general condition of information technology in business is healthy in terms of investment in advanced technologies. It also seems healthy when the most successful applications are reviewed. However, the average company is not able to make technology support today's work. While the unit costs of technology decrease, the investments increase and the returns are increasingly questionable. Application development backlogs are growing despite the increasing capabilities of new equipment and software. The various systems do not work very well with each other. In some industries, new systems development has all but stopped completely. Banking, for example, is already an information-based activity, but there has been little new systems development in the industry in the last ten years. There are almost no companies that have demonstrated the ability to take full advantage of the state of the art in information technology. This would seem to be due to the inability to relate technology support directly to the activities making up business processes.

Paradoxically, there is no doubt that most businesses could not survive without technology in general and information technology in particular. There is some justification in the view that business is only evolving as technological advancement allows it to. All the visible successes and problems attributable to the use of technology in business points to one conclusion: technology has great potential, but it is difficult to use effectively. Re-engineering addresses these problems by designing effective use of technology directly into revised business processes. It is accordingly strongly recommended that technology support be specified during the re-engineering project.

Defining Technology Support in Re-Engineering Projects

The re-engineering project is the best place to define the need for information and information technology support, as well as the use of new noninformation technologies. The places at which technology can be used will be apparent when the Business Activity Maps (BAMs) are created for the current business processes, if they are analyzed by someone who is computer management literate.

Computer Management Literacy was developed by the authors (*Working Effectively with the Information Systems Department*), videotape, Naperville IL: Deltak, now Applied Learning, Inc., 1985) to give business managers the background required to manage their use of information technology. We have found that business managers do not needs computer literacy for their work, except to use their own personal computers effectively. They do, however, need a different sort of knowledge related to automation to maximize the use of their companies information technology capabilities. The relationship between computer literacy and computer management literacy is shown in Fig. 8.1.

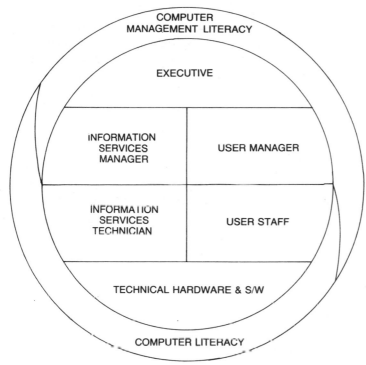

Figure 8.1. Computer literacy vs. computer management literacy.

Computer Management Literacy contains the following four main areas of knowledge:

1. *Capabilities*: What each type of information technology product can do for the business. Knowing the specific functional capabilities of mainframes, minicomputers, PCs, communications systems, storage systems, and others is required for management decisions regarding the use of technology in business situations.

2. *Limitations*: Least discussed, but most important, are the limitations of the technologies. These are the extent of the capabilities of the technologies: when they run out of capacity, what functions they cannot do well, which cooperation with other technologies is difficult for them, and which difficulties will be experienced by staff in using them.

3. *Costs*: What is the full cost of buying, installing, learning, using, maintaining, and supporting the technologies?

4. *Methods*: The methods by which the technologies are used include systems development methodologies, data center management approaches, cost accounting and chargeback methods, service level management methods, local area network administration, and a handful of others. Managers do not need to know the details of these methods, but they do need to know the basic structures and objectives of the ones related to any of the technologies that may be used.

Computer Management Literacy and business process re-engineering, working together, are the beginnings of the effective and controlled use of information technology. The Positioning activities will provide a good picture of the current technology support, in which the issues related to current information systems will be highlighted. At this point in the effort, it is recommended that an information technology expert be added to the re-engineering project team.

The subsequent analysis and development of effective technology plans will take place as part of the re-engineering project. The steps are as follows:

1. Assess current information services support and office technology related to the processes being studied.

2 Find tasks and subprocesses that have an especially critical requirement for information and information technology.

3 Determine how to improve the business processes by using information technology and also how to get the data required to make decisions to the critical points in the process.

4 Define the information technology architecture, technology support, and information systems that are needed.

5 Implement the required technologies, using the re-engineering models as a basis.

6. Use the re-engineering models on a routine basis to control the flow of information in the corporation.

The details of these activities are now described.

Assessing Current Information Services and Office Technology

The first step in the technology side of business re-engineering is the evaluation of current information and office technologies. This evaluation is based on information gathered during the Positioning effort. The objective is to establish a baseline for a redesigned technology environment, not to assess the performance of the current support or the adequacy of the information services staff. Much of this work may therefore be conducted by the information services staff, if their technical knowledge is fully up-to-date.

Assessing the Documentation

Most medium and large companies rely on computer support from a combination of internally written and purchased programs. Generally, the purchased programs will have acceptable, but rarely good, documentation related to their use. Documentation for applications systems that have been programmed within the company will vary widely. However, in most cases there will at least be basic documents, such as functional descriptions and data definitions available. Even the most rudimentary documentation (for example, file record layouts) may be useful to the re-engineering project. If systems have been written using standard process descriptions, they may provide very valuable input to the project's files. Certain computer-aided systems engineering (CASE) tools may even provide direct input into any automated re-engineering tools being used.

The documentation of the current technology may include any of the following.

Process charts: Any of many types of charts of the business processes, such as Warnier-Orr diagrams, data flow diagrams, or the Business Activity Maps of Relational Systems Development (see Chap. 5). These charts are found in connection with only a small proportion of systems, having come into use recently. They have also been less popular among systems developers than data relationship charts, the process side being left to the programs. Process charts are, however, critical to reengineering.

Data relationship diagrams: The charts that show the relationships of the data used by the corporation. When it is available, this information will be provided in the formats of either the information systems planning (ISP) or Peter Chen's entity relationship diagrams (ERD). The use of these charting tools is quite common for systems built in the last five years, especially when CASE tools have been used. These charts are only moderately helpful for re-engineering, however. This is due to the nature of data relationships: there is not much difference between the data relationships of one company and another. It is in the business processes that differentiation and therefore competitive advantage are to be gained. Therefore, data relationships are more important when designing information systems than when re-engineering.

User documentation: The instructions intended for users of the information services systems. This documentation gives fairly direct clues to the conduct of the business processes being supported, and are often useful for that reason.

Program design charts and programs: Program designs in any of several formats, such as Yourdon, James Martin's action diagrams, RSD's Relational Diagrams, structured pseudocode or the highest-level source language. Because most existing application systems have been written without the benefit of re-engineering guidance, programs do not generally match business processes. This reduces their value considerably. The value of most programs to re-engineering projects is that they document the most detailed level of key business rules.

All the documents of these types, which contain information pertinent to the areas of the business being studied, should be reviewed. Unless they are maintained in some automated form, they should be assumed to be out-of-date. In fact, unless the maintenance of the computer systems is conducted by modifying the design documents and then using CASE tools to generate the systems directly from them, it is unlikely that the design documents will be current. Unfortunately, the ability to maintain systems at the design level is not fully implemented in any CASE environment as yet.

As the documentation is reviewed, it may be augmented where there are significant shortfalls in available information. In some extreme cases, the current documentation may be completely discarded, and the information gathered by interview and from other corporate sources. Although it may be advantageous to ask information systems technicians to extract information from current systems that are poorly documented, it is seldom practical to redocument existing systems.

Assessing Current Technology Components

The next step is to survey the current hardware and software. Although the scope of the project's interest in software is limited to the systems supporting the processes being re-engineered, all the hardware used by the corporation is a potential concern.

The highest level of information technology design is the technology architecture. All companies have them, but technology architectures are seldom formalized. Therefore, a description of the corporate information technology structure may not be available when the first Positioning efforts are initiated. As one of the objectives of Positioning is the definition of this architecture, it will be either created or validated in each effort.

The architecture will describe the main categories of information technology and their relationships to each other. It will be designed by an expert to take fullest advantage of existing technology and to provide for the most productive movement to newer technologies as they become useful. Although the work that goes into designing an architecture may involve complex technical issues, the results should be simple and understandable. One of the most important functions of the architecture is to establish the approach for all parts of the company to share data and to interoperate with each other. This, of course, implies that the company's use of data must influence the technology architecture.

An example of the design of a corporate technology architecture is shown in Fig. 8.2. The technician first drew a chart of what was found in the company, which was not designed but just grew. Nonetheless, there was an implicit architecture. The issues raised by the implicit architecture were common ones in business. There was no strategic platform for the construction of client/server applications. These are applications built on both the mainframe and personal computers. Another issue was PC operating systems; the company could not decide whether or not to convert to Microsoft Windows or stay with MS/DOS, a common issue in business today. The ambitions of PC users seemed to conflict with the methods of the technicians responsible for the central transaction systems, such as the payroll, ordering, and accounting systems. PC users, however, were very interested in getting data from these mainframe systems.

The technical expert designed the architecture shown in the second chart in Fig. 8.2. The new architecture is based on the use of distributed databases and client/server applications systems. IBM's OS/2 was selected as the PC operating system that provided full support of existing PC systems and also the most strategic path for incorporating future technology improvements. The mainframe system, which is less easily changed, remained much the same. This new architecture will provide for the selection of technology products, building new applications systems, converging the interests of PC and mainframe users, and providing effective network services. This architecture is only an example, of course. Other selections will be more appropriate for other circumstances.

Technology architectures may always be represented graphically in some way. If an architecture is discussed, and it is stated as being logical, or conceptual, rather than physical, or if it cannot be depicted, then there is no architecture. It is also a good idea to limit the chart of overall architecture to a single page.

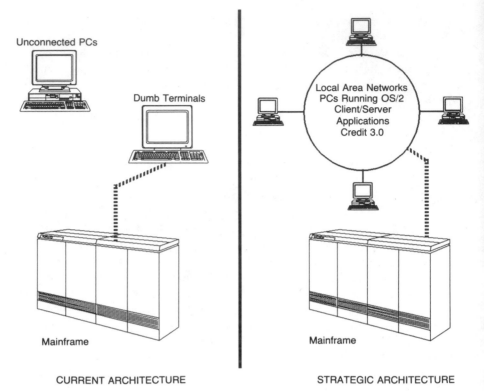

Figure 8.2. Old and new technology architectures.

Assessing Data Quality

One of the most important foundations for information technology support is the data itself. The quality of the data is obviously of primary importance. Any investment in information technology may have no payoff—or, worse produce a liability—if the data is bad.

The data for which quality assessments should be made include not only computerized data, but all data that is used by the project's business processes. This data may be found on forms, in the notes of the marketing department, in automated production equipment, in some isolated personal computers, and in many other places. The quality of the data found in some places may be difficult to determine. Furthermore, data quality depends on characteristics other than accuracy, such as timing and organization.

Finding data problems should begin with the consumers of the data. The managers and staff who use the information in PCs and reports are keenly aware of data problems. In some cases, the systems themselves will provide additional help in evaluating data quality, by providing special checks that result in data quality reports.

Correcting data problems is generally more difficult than just changing the way that source information is transferred. Data quality assurance presents the same challenges as any other quality improvement program. Thus data quality improvement may require basic changes to the way the data is gathered, changes in the technical aspects of data movement, and altered (or new) databases for storage and retrieval. These activities affect the business processes themselves and should influence the re-engineering strategy. For example, inferential tests should be built into the data-gathering systems. These tests compare critical data elements against the values of other data elements, not simply against predetermined parameters. An example of an inferential test is if the number of orders for a month is above average, the amount of goods shipped should also be above average. If this relationship is not proven to be true in a monthly data test, the system should flag the data as being in error.

Due to the growing importance of information to the company, data quality is an extremely important issue and merits attention, indeed more attention than is usually given to it.

Assessing Work in Progress

The final area of current technology to be assessed is the work being done to improve the company's information services: work in progress, work that is in the planning stage, and work that has been proposed but not yet planned. This last category, which includes the ubiquitous and menacing applications and maintenance backlogs, is especially interesting.

If any systems, hardware or software, are being implemented, they will require priority attention, of course. An early decision must be made as to whether to continue these efforts. It is also possible to alter efforts in progress to conform to the re-engineering project if there are known conflicts. It is generally not feasible, however, to suspend information services efforts until more is known about the re-engineering effort. The decision to proceed or stop should be made and then carried out; it will cause nothing but wasted effort to continue these efforts and then cancel them at a later date. Often a new information system will seem to be an unprofitable investment in the existing processes, but the new system may in fact help the change process by putting some processes or data under better control. New systems tend to be much easier to change than old ones.

The backlogs, the requests for systems that cannot be provided, are to be scrutinized as a rich source of ideas for improvement. The assumption is often made that backlogs are made up of ideas that would not give value for money. Most companies readily implement high-impact, low-cost proposals. However, many of the proposals in the backlog have not been funded for other reasons. In some cases the cost of the proposed information services changes was prohibitive only because of the construction or condition of the existing computer systems. If these ideas are built into the re-engineering effort, they may yield significant benefits at almost no cost.

Finding Information- and Technology-Critical Areas

As the positioning and re-engineering efforts progress, Business Activity Maps (BAMs) and Relational Diagrams will reveal both requirements and opportunities for the use of information and technology. Analytical efforts can be conducted to determine the best ways to satisfy the requirements and to take advantage of the opportunities. The results will provide the basis for information technology support and also influence the design of the business processes.

Using the Business Activity Maps

The Business Activity Maps are the first place to look for points in the business processes where information and technology will be important. The very highest-level BAMs may even identify subactivities that have names similar to those of the company's major information systems, suggesting that these subactivities or even the entire process will be supported by a large automated system. However, the detailed-level BAMs will provide much more reliable information on the nature of the information systems that will best support the work process. Because technology can, and should, influence the design of business processes, the technology analysis should be done at the same time as the BAMs are being drawn, as part of the re-engineering effort.

Before looking at a BAM, the change analyst must know what support information and technology can provide. The help that technology can provide to the business process is, as already described:

1. Increased speed.

2. Storing and retrieving.

3. Communicating.

4. Controlling process tasks and improving quality.

5. Monitoring.

6. Supporting decision making.

7. Manufacturing and delivering services to customers.

8. Supporting work functions.

Identifying opportunities to use technology assistance will be among the analyst's primary goals. Additionally, the analyst will identify the need for specific information at each point in the BAM. The two classes of technology requirements (needs and opportunities) identified in the BAM should, of course, be compatible with each other, and may be addressed together by technology solutions. These two classes of technology, along with the application technology, are the technology points of the effort.

In practice, it is usually easier for the analyst to begin with the information requirement points, because they are the most obvious. Information, as it is used in business, comes in several forms. The first form is *operational* information, which is information whose content does not determine process flow, but that must be manipulated in some way. Operational data is like production materials to a process: it contributes to the product, but it is not used for decisions. The information that makes up bills is operational information to the billing process, for example. The billing process puts this data on a billing form, but does not look at it and take action based on it. The data is simply placed on the bill and mailed out with it.

The second form of information is used for *decisions*. For fairly well defined processes, these decisions will be clear and will have clear criteria, but some business processes, such as marketing and new product development, may have what are called "fuzzy" decisions requiring ad-hoc data. Closely related to this decision data is *control* data, which does not decide yes or no, but how much or how many. Control data is used to control process such as manufacturing, assembly, supply ordering, and shipping.

Figure 8.3 shows an example of a Business Activity Map with an analyst's notes identifying information and technology critical points. These notes identify the potential uses of technology as well as the information entities used by each task in the process.

Improving Business Processes Using Technology

When the basic technology and information points in the BAM have been identified, the most important step takes place. The analyst suggests changes to the process that can be made by using new technology implementations. These changes should simplify the process, improve the

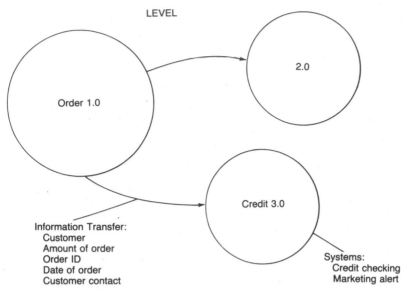

Figure 8.3. Business Activity Map showing information and technology.

quality of products or services, reduce the cost of the process, decrease delays in the process, or make some other measurable improvement. Because the BAM includes measurements of the process's costs and performance, it should be possible to assess the improvement directly. This solves one of the most significant problems posed by the use of new technology in business—determining the contribution of the technology investment. The method is simplicity itself: the costs and benefits of the process after the technology improvement are compared with the costs and benefits either of the current process or of the process as it is best configured without the technology improvement. Alternate technology solutions can be compared using this same methodology. The methods used to estimate the costs and the contribution to the profitability of processes using the Business Activity Maps are discussed in Chap. 7.

Improvements can occur in both operations and administration processes. Each of the ways that information technology can support business (just described) provides a potential for improvements, but only a technical expert working with the re-engineering team can judge what available information technologies can do to help. The best places to look for improvements are where redundant information activities occur and where there are obvious delays resulting from information not being where it is needed. A good example is a process in which a document is being prepared. The document goes through several reviews, each of which results in changes, reprinting, and redistribution. If a technology improvement is made, the document may be placed in a network access file so that all of the reviewers can see it immediately. Also, the reviewers can see each others' comments. The changes are also available as soon as they are keyed in. The entire process can be shortened considerably, and even the cost of paper and distribution reduced, by using technology.

Another opportunity is increased efficiency by getting to decision points data that is not currently available to them. This situation will be seen on the BAM as tasks that direct the flow of work to where there is certain decision information, instead of moving the information into the flow of work. For example, when orders are sent for clearance to a credit group, the process (or flow of work) is moving to the data. Instead, the credit clearance can be entered into a technology system (or can even be determined by the technology in advance), so that the order need never leave the order department.

The re-engineering project may model work processes containing many alternative technology solutions, and compare the resulting costs and benefits. The technologies proposed and modeled will not be limited to traditional systems that might be programmed by the company's information services department, but will also include office technology systems, procurable applications for both PCs and mainframes, local area networks, wide area networks, subscriptions to public information databases and news services, and, in fact, the full range of information technology tools available. The modeling process will enable the best technology solutions to be identified.

Defining Requirements for Information Systems and Technology Tools

Re-engineering should always be used to define requirements for information systems. A basic assumption of Dynamic Business Re-engineering is that business processes and their technology support cannot be designed separately. Furthermore re-engineering does some of the work involved in developing or buying information systems. The requirements analysis, the traditional first phase of information systems development, is mostly accomplished by the re-engineering project. The requirements for technology support will be detailed in the re-engineering design.

The results of the Dynamic Business Re-engineering design effort will include the complete definition of the technology support requirements for the process. To complete the systems described in the re-engineering design, the information services systems analysts now need to perform the more technical tasks of developing the final levels of detail specification, defining the information architecture for the process, selecting the technology components that will be used, and then programming and implementing them.

Defining the Information Architecture

The Dynamic Business Re-engineering charts show the basic data that is needed by the processes, where it is needed, and by whom. The next step is to determine how the technology will organize the data: the information architecture. This entire function will soon be automated, but today one of the most significant limitations of automated database design is that it will not create a distributed data definition. In the unlikely circumstance that it is most efficient to keep all the information in one database system (which implies that all the information is to be automated), technology products are available that will accept the information in the re-engineering charts as input and generate a computer database.

In general, an analyst will use the re-engineering charts, annotated with information uses, to establish an information architecture. Dynamic Business Re-engineering, using the methods of Relational Systems Development, uses the RSD Hierarchical Interface Relationship Diagram (HIRD) to lay out the basic information, systems, and technology architectures (Fig. 8.4). To create the HIRD, the analyst examines the re-engineering charts and places stores of data as close to the work as possible. The requirements to share data are then satisfied by either remote access or by moving the data to central stores. Data integrity and backup are also taken into consideration in locating the data.

This ability to distribute data and computer processing is changing the way systems analysts and designers look at systems. A system in the past was a group of programs that were used to support a common business function. In the context of today's technology, a system is more likely to be defined as a database, or pool of data, residing in a technical environment, with all of the programs associated with it.

Defining the Technology Architecture

The HIRD also indicates the technology platforms (the type of computing equipment) where the data resides, and from which the systems are to be constructed. This architecture is more specific than the corporate

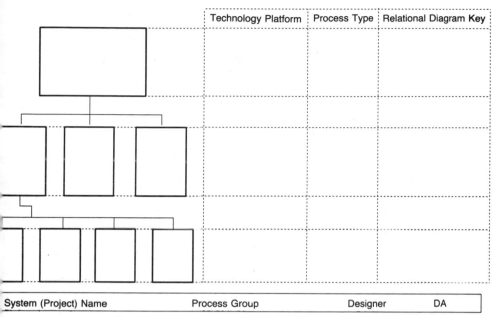

Figure 8.4. Blank HIRD.

technology architecture. Whereas the corporate architecture may indicate a choice, the HIRD makes specific selections. The two charts should not be in conflict, of course. If a very new requirement or a significant technological advance necessitates a different architecture for the process from the one designed for the corporation, then the corporate architecture should be formally changed.

The HIRD will indicate the specific technology tools, hardware and software, that is to be used at each level in the technology structure and on each computer. There is no formula for this determination. As the technology advances, the choices increase. However, the corporate architecture, designed to assure compatibility among all the company's technology components and the ability to change as the state of the art changes, will narrow the choice for each process HIRD. If very new technology capabilities are needed, they will have been proposed during the re-engineering process, and these will not generally pose problems of choice for the construction of HIRDs. In some cases there will be no technical reason to choose one technology over another and, in these cases only, competitive bids may determine the choice. In general, competitive procurement is not a good approach for technology acquisitions, despite the widespread belief to the contrary.

Buy or Build?

The definition of the technology systems in the re-engineering process will also help to determine whether to build or buy a system. Obviously, the question can arise only when there is a commercially available system that satisfies an applications requirement. Also, no company will consider building its own technical software, such as operating systems or word processing software.

In those cases where there is a choice to be made, the results of re-engineering can help to make it in two ways. First, the re-engineering charts will show the requirements of the system, so that they can be compared to the off-the-shelf software. Dynamic Business Re-engineering provides not only the functional requirements for the system, but also the operational requirements: how the system is to be used in production. This will provide a clearer assessment of the candidate software systems. Second, the re-engineering designs will show the contribution the software is expected to make. If this in any way involves giving the company a competitive advantage, the ability of purchased software will be questionable. If anyone can buy it, what competitive advantage will there be? Consideration of competitive advantage has not been applied to the purchase of software in the past, but today it may be the most important factor.

For some work common among businesses, purchased applications software will continue to be chosen. A good example is basic accounting software. However, the trend toward the increased use of purchased software that began in the 1970s seems to have run its full course. With new developments bringing the costs of custom software down, with competitive advantage becoming a growing concern and with re-engineering doing much of the design, building custom software will become the most common choice in the next decade.

Determining Presentation Requirements

In recent years, systems designers have become aware of the importance of presentation. The new desktop technologies have made graphical presentation capabilities available to the analyst and systems designer. These graphical capabilities have revolutionized the way people are able to interact with computers and have resulted in systems that are easier to use and more readily accepted. The choice of presentation technology may thus have an impact on the design of the processes and will certainly influence the implementation of the new system. The presentation methods available include old-fashioned printed output, terminals or terminal emulation, presentations using personal computers, and presentations made directly by equipment and computers performing business process functions.

The most difficult choice in designing computer support today is whether to use PCs as dumb terminals or as intelligent computers. The selection will be predetermined if the information architecture places some of the databases on the PCs. However, when the PC is not used to store data, there may still be good reasons to use some of the PC's computing capability for presentation management. A growing number of good PC tools make this possible. The determining factors will be the complexity of the interface between the computers and the staff operating them, and the response capabilities of the network. If, for example, the network is slow due to the need to use long distance connections, the PC can perform enough of the data presentation function to keep the operator from waiting.

The design of presentation includes provisional drafting of actual screen formats and discussing them with the re-engineering team. The screen and report formats will provide detailed specifications for the new technology systems, and will also give the team a better sense of the new business process, from the viewpoint of the staff who will actually do the work. It is also possible to use the technique of prototyping to help design the screens and test them; this is a subject best left to the systems development technicians.

Implementation

When the re-engineering project hands its designs over to the information services department, much, if not all, of the systems analysis and design will have been completed. The system will be ready for implementation. Without re-engineering, the first phases of systems development efforts (customarily requirements analysis, systems analysis, and systems design) were the second largest portion of the cost of information systems over their life cycles, behind only maintenance, which typically accounts for 80 percent. Thus the remaining tasks, between these first phases and maintenance, which include programming, testing, and installation, never accounted for the great amount of effort and resource that they were thought to. When the systems development effort is reduced to these three phases, it becomes manageable and affordable. Risk is also reduced, almost to the point of elimination.

The Programming Phase

Using the input from re-engineering, the additional detail added by the RSD design methods, and the automated support provided by a number of CASE tools, the programming phase is almost fully automated. The CASE tools provide the factors needed to construct a program from the RSD and Dynamic Business Re-engineering charts and tables. The Business Activity Maps contain the overall logic and flow of the system. The Relational Diagrams contain the detailed logic of system, the actions that users will take in operating the system, and the scenarios to be tested. The positioning tables and matrices contain the business rules that will be translated into program code. The data relationship diagrams contain the data definitions and relations required to create databases. The Hierarchical Interface Relationship Diagrams describe the division of the systems among the technology platforms (PCs, mainframes, etc.). Finally, the screen and report definitions contain the detailed formats for these input and output vehicles.

Having all the required definitions, the systems development team is able to very quickly generate or code the programs required. The technology tools currently available are not able to generate 100 percent of the code for any technology platform or database system, and for some technologies no code at all may be generated. However, this capability is increasing. The code that cannot be generated can be written with relative ease from the RSD charts. Programming using structure charts for definition is an established approach, and should be used in all programming efforts even when no CASE tools are used and no re-engineering or positioning has been done in preparation. This work should be well within the capabilities of most information services departments.

The Testing and Installation Phases

When systems are designed to support re-engineered business processes, the entire process goes through testing. Testing new systems under Dynamic Business Re-engineering is also much improved over traditional systems development methods. The systems are part of the process test, with the test scenarios set in accordance with the functions and timing of the re-engineered business process. The increased levels of detailed knowledge developed by re-engineering remove much of the potential for surprises during the testing of the systems, and provide the user with a complete framework in which to evaluate the systems.

Re-engineering does not, however, shorten the testing of information systems. Because more is tested, the testing may take longer. This is time well spent. The tests of new processes will not only assure the proper functioning of the new systems, but also the timings, the response of the system under full production workload, and the systems' ability to realize the full benefits predicted for them.

Installation, like testing, is facilitated by re-engineering. The plan for phasing over to the new business process will incorporate the installation of the new information systems. The users will have a full set of new operating procedures. The re-engineering environment addresses the most serious problem associated with installations in the past: the degree to which the systems change the work of the users. In a re-engineering project, these changes in work are the purpose of the effort. They are well planned and managed and are fully introduced to the workforce in advance of the installation of the systems.

Systems Maintenance Under Dynamic Business Re-engineering

Since maintenance has been the phase of the life cycle of systems in which most of the cost has been expended, it is the phase that should receive the greatest attention when trying to achieve efficiencies. The main advantages of CASE tools are indeed realized in this phase. If systems could be completely generated from their structured design charts by CASE tools, maintenance could be performed on the design charts, saving enormous amounts of effort. CASE theory also states that if maintenance is conducted on the design charts (and only if it is), the charts will be kept up-to-date, eliminating one of maintenance's chief problems.

Re-engineering goes far beyond CASE in its impact on information systems maintenance. First, re-engineering, due to its deeper definition of requirements and more effective testing, reduces the component of maintenance devoted to fixing problems to a minimum. This effort is high in current systems only because they are unsystematically constructed and insufficiently tested. The second component of testing, the enhancement of systems, can be viewed in two ways: either highly facilitated or eliminated entirely. Re-engineering facilitates enhancement by providing an easy path to making fully controlled changes in the business processes and systems. This is another example of the reuse of the same models that were created for the re-engineering project to obtain systems enhancements. Each reuse of the models will benefit as much as the first use from all the tools of Dynamic Business Business Re-engineering.

A more interesting way to look at maintenance under re-engineering is that it may no longer exist. After the first exercise of positioning and re-engineering, the systems and the business processes that emerge will no longer have a life cycle in the traditional sense. Because re-engineering will be done on a routine basis, information systems will be re-engineered at the same time. They will always be new systems, never patched or added to. Because the costs of new systems are reduced by the improvements introduced by re-engineering and CASE to a fraction of what they have been, there can be a great increase in the amount of systems support and the use of old, outdated systems can be relegated to history. A multimillion-dollar investment in computer systems that took many years to develop can be replaced in a few months for a modest cost. This capability is another reason that purchased systems will decline. The in-house costs with which they compete will plummet.

Using Re-engineering Data Models in Production

An unexpected benefit of the Dynamic Business Re-engineering methodology is the use of the re-engineering models in the daily operation of the re-engineered business processes. The process models themselves are used for process control and as standards of performance, as is described in Chap. 10. The data models developed in re-engineering and subsequently in systems development, using Relational Systems Development methods, can support the management of the data used to control business processes.

The data models tell where the data is located, so that ad-hoc data needs, which occur throughout the company, can be fulfilled more easily than at present. The same models show the expected flow of the data, so that communications can be planned and monitored under the control of real business priorities. These capabilities do not exist today. The models can also be used as the basis for future re-engineering and systems work in other parts of the business, where similar work may be done or similar data may be used.

9
Re-engineering Human Resources

The human side of enterprise is second to no other factor in importance. In a business that employs a workforce of any size, success must depend upon their performance. While it is rare to find an organization that does not employ a few people who contribute less than they take, none can tolerate more than a very small percentage of such and survive. Also, there are no cases where even one poor performer does not to some extent reduce the performance of the group. If business relies on performance, then so does re-engineering. The re-engineered business process, if it is designed to be more efficient than the old, may be even more dependent on everyone in the company performing the work for which he or she is responsible.

There are more assumptions about personnel management than organizational issues, and they are adhered to with more emotion. Management assumes that each position is important, and that among the most important responsibilities of a manager is the challenge of getting the most out of each person. However, managers also assume that it is impossible to get 100 percent from either the workforce or any individual worker on a consistent basis. This is often viewed as hypocritical by the workforce: especially when management says that the workforce is the most important asset of the company, but then treats it as though it had no value at all. Many managers seem to try to impress their superiors by displaying tough attitudes, especially the willingness to fire workers for any reason that appears to benefit the company. Corporate personnel administration is seen as villainous by both workers and managers, taking the side of management in matters related to compensation, and seeming to take the side of workers in matters related to discipline. The relationship between labor and management is confused, and the theories that address this important area are conflicting.

The industrial revolution began with the view that the operator of a machine did not require great skill and would not be treated with the respect or be given the compensation due to a craftsperson. The culmination of the industrial revolution was the development of the assembly line early in this century. The early efforts of management science, such as Frederick Taylor's work, reinforced the view of the workforce as part of a machine. Industrial engineering can be looked upon as directing the worker in the last detail of the work, leaving absolutely nothing to individual volition. This is, of course, not the actual attitude of most modern professional industrial engineers, but their work appears to lead in that direction. The design of many industrial processes took the complete definition of work as a goal, and most service and process work flow is based on the same assumption. If this assumption is carried to its furthest, the ideal business would be staffed by nothing but robots. Most of the experience of managers and the history of the development of business has done nothing to contradict these assumptions. Robots work very well in factories; people can usually be counted on to provide some sort of difficulty.

In recent years, the mechanistic assumptions of ideal work functions were questioned. The use of action research and organization development produced some astonishing success. Organization development leans away from complete job definition, toward the theory that, if management adjusts the environment properly, the workforce will define the work for themselves. Few managers will accept this premise completely, and indeed organization development does not suggest that work definition be completely forgotten. However, challenging the job definition school of thought has been a rich source of new ideas. Once the old mechanistic assumptions were questioned, new horizons of management were opened. Perhaps the mechanistic ideal is not ideal at all. Perhaps the workforce can be trusted with much of the responsibility for job definition. If so, could not management itself be made more efficient, could it not get better results with less effort? Would changes then be much easier to implement? Organization development techniques for solving specific efficiency problems are now also providing the promise of more far-reaching benefits. The latest quality improvement methods do not rely on industrial engineering approaches, but upon organization development ideas.

Re-engineering has entered the business world more or less on the industrial engineering side of the management science conflict. Carried to its extreme, re-engineering can be made to define every motion of the work to be done. Does this imply that the potential gains of organization development and quality improvement are in conflict with re-engineering? The answer is that it is certainly capable of setting up a very detailed work process, completely ignoring the abilities of people, as well as trampling out any vestige of initiative on their part. Fortunately, it need not be done that way. Just as generations of industrial engineers have managed to temper their own logical and mechanical approach to work with human values and methods that encourage individual initiative and flexibility, so may the re-engineering project. Unfortunately, most re-engineering projects have not been staffed with either industrial engineers or organization development specialists, and have ignored these issues. Providing a good balance between work mechanisms and teamwork is possible only if the need to do so is known.

The overall recommendation made in this chapter may be described as *plan from the top down and implement from the bottom up*. This advice is applicable for almost all types of change projects.

Personnel Issues and Re-engineering

A full range of personnel issues can arise from a re-engineering project. The need for recruitment, training, job regrading, transfer, restructuring parts of the organization, outplacement, and counseling may occur. The only regular human resources department functions that will not normally be expected are disciplinary actions—not, that is, unless the project is badly mismanaged.

The importance of human resources to the success of the re-engineering project make this an area that should receive attention from the beginning of the project. The participation of the personnel department can help to identify problems while there is time to solve them without delaying the project. It can also provide information concerning staff and help to re-design the business processes. Obviously, it is highly desirable for the new processes to be compatible with corporate personnel policies. It is surprising how often plans laid with the help of the personnel administration staff fit easily into corporate policy, while those that exclude them almost never do so. If the company has organization development experts in the human resources department, they can be very helpful in the design of the business process and in addressing implementation issues.

The issues that the re-engineering project will face may not surface until a new business process design has been drafted. At that time, the need to translate the skills of the current staff to the new work requirements will be apparent. This issue is the lead item in a formidable list of others.

Human Capital—the Most Valuable Resource

When the re-engineering project begins to address staffing issues, the value of human capital will take on a new meaning. An awareness will emerge that the key to a successful re-engineering effort is the workforce. The staff will be charged with making the new operation work; no design can be implemented if the staff fight it. Any ivory tower isolation that has found its way into the re-engineering team begins to vanish with this realization.

The issues that the re-engineering project must address to staff a new process can be intimidating. The result must be people, with the right skills, doing the newly defined work. The project must also deliver any streamlining of the staff that has been promised, which may require cutbacks. These two objectives result in a long list of activities:

1. Defining positions and skills.
2. Defining a new organization.
3. Redeployment.
4. Training and retraining.
5. Regrading.
6. Outplacement (if required).
7. Implementing changes.

During all this, the morale of the workforce should be kept at a high level. These tasks seem almost overwhelming, and many change projects have been overwhelmed by them.

Fortunately, there are ways of solving the staff issues and managing the implementation of re-engineering projects that turn the difficulties in dealing with people into advantages. The workers in business may not be able to carry out functions as well as machines, but they can do some of their own implementation work. They can, in fact, carry some of the burden of their own management, and can help to design and implement the new work processes, organization, training programs, and adminis-trative processes. In doing so, they may solve their own morale and motivation problems. It is even possible for the workforce to exceed the design requirements specified by the re-engineering project. That is what makes human capital the most valuable resource.

Realizing Staff Cost Reductions

Before discussing the details of how all the personnel work can be done, the issue of staff costs conservation should be examined. This is often considered the central issue of re-engineering and its principal motivation. In most cases, however, the management of a business does not turn to mass staff reductions unless the need is urgent and there are no alternatives.

Re-engineering may indeed seem an ideal approach to staff reduction. Any business process that increases its efficiency should achieve either higher output with the existing staff or the same output with less staff. Quality should not suffer in any case. However, experience has shown that the problems that occur when other forms of change projects (the common reorganization, for example) try to incorporate reductions in force, also affect re-engineering. These problems include difficulty in retaining the right staff, complete loss of trust in management, loss of institutional knowledge, very high outplacement costs, adverse impact on productivity, and a definite tendency of the affected areas of the business toward restaffing themselves. Another common problem is that reducing the size of the staff may require more time than the rest of the change project, and may delay the implementation of other beneficial changes. In many cases, the short-term benefit is reduced by implementation costs and the long-term benefit never materializes.

A more satisfactory strategic objective than immediate reductions may be downsizing over a given time span, using normal staff attrition in place of outplacement. This policy helps to control growth in the size of the staff while increasing motivation. There is a planned and enforceable reduction in recruitment costs and the only cost associated with ongoing staff are those that would have been incurred had no changes been made. As staff reassignment is often an integral part of this approach, training costs are likely to increase; but training is much less expensive than recruitment. If this method is used, both the pressure of outplacement and the usual resistance to the change project associated with downsizing can be avoided.

If re-engineering is specifically prohibited from reducing the size of the workforce beyond what attrition can provide, the project will be under considerable pressure to find other ways to improve corporate profit. The efficiency improvements will likely be directed into operational streamlining and waste reduction. This approach shifts the emphasis on cost savings from a negative to a positive as it promotes a common goal. Effort is thus invested in profit-making initiatives that will most likely provide a much higher long-term benefit.

Defining New Jobs

The re-engineering efforts will have produced a new set of business processes, which will require staffing changes. One of the problems re-engineering set out to solve was the fragmentation of positions that resulted from attempts to reduce effort without re-engineering. It is therefore assumed that the new processes will have been designed so that whole positions could be devoted to the newly defined work. This, however, depends on how the process's work is divided into individual work assignments.

There are two or three approaches to organizing the work into individual jobs, units, sections, and departments. The first is to apply re-engineering design techniques to a very low level and then to group tasks together into jobs, jobs into units, and so on. The second is to stop the re-engineering design at a higher level and let the working level staff design the rest themselves. This is a participative management approach, the type that organizational development would recommend. An even bolder approach is to assess the work in a certain area, determine the basic skills that are needed and begin to recruit members for the re-engineering change team from the current workforce. Once the effort is completed, the team members will become the staff assigned to perform the work they re-engineered.

If the first approach is taken, the methodology is simply that of re-engineering, as described in Chaps. 5, 6, and 7. There are no special tools required. Even in the precise environment of this approach, some flexibility exists. The new jobs may be defined to approximate the old job descriptions, for example, or, in fewer cases, may be designed de novo. Using old position descriptions saves effort and will give some workers a more familiar basis. The benefit of writing new job descriptions is that they can usually be improved. The average job description is written to support job grading and to provide a basis for performance reviews. They are not intended to tell anyone how to do the work, which is one of the reasons they are of so little use as input to re-engineering. Re-engineering can help to create more effective job descriptions by providing a clear explanation of the work, the work's relationship to the business and quantitative standards for performance. These factors, the job's linkages to other jobs, and the objectives of the work are far more important than the lists of tasks and sets of qualifications generally found in job descriptions.

The methods used for organization development approaches, in which most teamwork schemes are included, are described later in this chapter.

Retraining and Redeployment

In re-engineering projects, retraining is used as an alternative to outplacing one employee and hiring another. Retraining and redeployment are used sparingly in comparison to outplacement, but they can be much more useful. Operating in the change paradigm, there will be many opportunities to redeploy staff. The need for training in these circumstances may differ from current corporate training programs in several important respects.

Training is currently given to raise an employee from one level of capability to another. Since most business training is purchased from sources outside the companies that use it, training design is based on the availability of materials. Because of the necessary progression of skills this implies, designing this type of training program generally requires a significant effort. The re-engineering process supports training design by providing detailed work process requirements: the specific background and skill requirements of each position are related to the activities of the new process.

In most cases, the process charts and the Relational Diagrams provide detailed documentation that has proven to be very useful in training. Workers can learn about the overall process they will be part of, as well as the details of their individual jobs, directly from these charts. They can also develop an understanding of the timing considerations and the importance of the work they will be doing. This understanding helps foster a sense of pride and self-worth, both important characteristics in the quest for quality.

Organization Development Approaches

Organization development (OD) is a new force in business management. It can trace its roots to the 1930s, but it was not generally practiced until much more recently. However, it has been used long enough to have proven its effectiveness. In some situations it is the only approach to change that makes sense, but re-engineering and OD are not natural allies. The theoretical basis of OD and re-engineering can be thought of as opposite ends of a spectrum of business management science. However, the two meet and cooperate very well when the change paradigm is added to re-engineering. This ability to cooperate has allowed OD to be used as one of the tools on which Dynamic Business Re-engineering is based.

What Is Organization Development?

Organization development is a collection of methods that help organizations improve themselves, mostly through change management. OD is a very broad field, concerned with industrial engineering, industrial psychology, training, and team building. As all organizational change, behavior, and performance factors are of concern in OD, the methods that fall at least loosely under the organization development umbrella number in the hundreds. Because some of these methods have acquired the stigma of faddishness in the opinion of many business people, it is difficult to recommend organization development without discussing the particular parts of this vast field that are of interest.

Organization development traces its history to several sources. One of these is the technique called action research, which is occasionally found as a separate discipline today. The term was first used in describing a problem-solving method developed by John Collier, who was Commissioner for Indian Affairs between 1933 and 1945. Collier used a joint action team approach to help solve problems involving race relations and other complexities that were far removed from the capabilities of industrial engineering (J. Collier, "United States Indian Administration as a Laboratory of Ethnic Relations," *Social Research*, Vol. 12, May 1945, pp. 275–276). This early work was conducted in the mid-1930s. Another initiator, also using the phrase "action research," was Kurt Lewin, a German psychologist, whose interests lay in the pragmatic application of science to social problems. Another source of OD is the work done at the National Training Laboratory, which developed the concepts of T-groups and sensitivity training. Yet another origin of organization development approaches was the Tavistock Institute of Human Relations in the United Kingdom.

Probably the best known OD success story was the work done for Volvo at its plant in Kalmar, Sweden. Volvo designed its new facility there to support an alternative to the assembly line. The plant opened in 1974, using a multiple team approach to car assembly that worked so well that it provided the basis for questioning many basic assumptions of business throughout the world. It also gave organization development a great deal of credibility.

Organization development has many approaches to managing change, no one of which is dominant. One of the most respected textbooks was written by Edgar F. Huse (*Organization Development and Change*, St. Paul: West Publishing Co., 1975). Huse describes organization development methods as having common underlying principles regarding people as individuals, as members of groups, and as members of organizations. Organization development tends to place more emphasis on behavior than most other industrial management fields, but its theory does not lack a systems, orientation. If there is a consistent theme in OD's view of systems it is a concern for interrelationships. In general, organization development is a very sophisticated approach to management, which is essential in dealing with very complex organizations, and increasingly attractive to the entire business community.

Using Organization Development in Re-engineering Projects

Can organization development, the antithesis of top-down structured, monolithic management techniques, support a re-engineering project, which may be the ultimate form of structured management? They can and should be used together; their opposite viewpoints enable each to overcome some of the weaknesses of the other. In merging these two approaches, the key benefits of both are sought. For re-engineering, the ability to determine the results of the project, assess its impact across the organization, and define support (such as information systems) must be preserved. From organization development, the commitment of the workforce, enhanced individual performance, and teamwork are to be gained.

The general approach to combining re-engineering and organization development is straightforward. It is consistent with the widely accepted philosophy that recommends plan from the top down, implement from the bottom up. The re-engineering design is carried out to a sufficient level of detail and then Organization Development takes over the implementation of the work flow. The three questions that arise immediately are:

1. What is a sufficient level of detail?

2. What organization development techniques should be applied?

3. How does this approach take over the work?

Although the answers to these questions vary, some guidance can be offered.

First, the level of detail that must be achieved in business process documentation before Organization Development can begin will vary based on corporate culture and the nature of the work being planned. The freer cultures and less mechanical processes will require less detail. OD offers direct support in these instances. Greater detail will not, however, prevent organization development from being effective, although some of the detail may require revision. Furthermore, complete detail will be required for information systems support. It is therefore better to err on the side of more detail than less.

Secondly, the selection of organization development techniques is highly dependant on the circumstances. The only two certainties are the use of teambuilding techniques (there are several of them) and job design.

Answering the third question, as to how organization development should be invoked, involves the organization of the entire re-engineering project. The project begins with the appointment of a change team, which is managed using the concept of an empowered, high-performing team. This team is staffed from the onset with experts in re-engineering, technology, management, and the business processes under study. As the project progresses, it is possible to add new members to the team: staff who will become the workforce for the re-engineered process. Toward the end of the re-engineering design work, an organization development facilitator can be added to the team. When the project is ready, the re-engineering and technology experts can go on to other work. The staff remaining on the project can then begin to plan the details of the work and actually begin to do the work as a team. During the Organization Development phase, the re-engineering tools will remain at the disposal of the team. These tools will continue to be available indefinitely to support production.

The working team may alter the process design within the scope of their work, but they must fulfill the expectations of senior management, which are clearly expressed in the Positioning and Re-engineering documents.

This approach may be used for new or highly changed processes, or for projects that tune existing processes, when no changes in the current staff are anticipated. For these less ambitious projects, the current staff should all be recruited for the change team, but on a part time basis, of course. This approach supports a smooth and productive interface between re-engineering and OD, and brings current process experience directly into the change team.

Obtaining Commitment from the Workforce

One of the chief advantages of organization development is that it develops commitment. This is undoubtedly one of the toughest problems facing re-engineering. Once a good new process design has been developed, by very systematic means, how does management go about convincing the workforce that the new design is a good one?

The OD approach is to involve the staff in the design work. Anyone who is among the designers of the new process can be expected to support it. There are some complications, but this method has been found to be very effective. The other advantage that re-engineering has is its ability to isolate the work of individuals and to show how the work of each contributes to the success of the process. This is usually motivating and, when it is not, the same process designs can provide standards against which the performance of each staff member can be measured.

Nontraditional Organizational Rationales

Re-engineering can be used to design work processes and organizations that look very much like current ones except for increased efficiency. However, these structures may not be the most effective ones in today's business climate. There are alternatives and re-engineering can be used to implement them.

The Need for New Approaches

The traditional organization has taken us a long way. With only a few embellishments, such as more effective project and quality control, the hierarchical organizational structure has won every war in history and has put human beings on the moon. The fact that alternative forms of organization are being tried at all is a clear indication of the force with which complexity and competition have assailed the business world.

The first indication that there was widespread discontent with the way things had been was the beginning of the trend toward increasing the span of management control. It had been accepted that the maximum number of staff that could be effectively managed by one person was between 7 and 10. Then the size of middle management came under executive scrutiny and the number has jumped to 20 or more. The former limitation was based on the number of close relationships that could be supported by one person. Apparently managers are no longer expected to establish close relationships. This indicates an entirely new use of the old hierarchy.

Although less popular, other initiatives have appeared that are intended to control the problems of hierarchies. The hierarchical structure has an inherent tendency toward growth. It will grow over time unless steps are taken to stop or reverse this tendency. Hierarchies also tend to depend highly on management. The quality of the work depends on the quality of the manager, so that, as the management chain grows longer, the probability of good work becomes very small. These problems have been dealt with in many ways, but recently there have been a number of attempts to confront them directly by using nontraditional structures.

The motivation for changing structure usually begins with the desire to improve competitiveness. The logic that leads to new structures often starts with the need to compete better, goes from there to the need to increase individual leverage and motivation, and ends up with a new organizational structure. The characteristics of the new structures are as follows:

1. Move the workforce closer to the customer.

2. Empower each worker: give each the authority to do more for the customer without asking management.

3. Shrink the decision chain: when decisions must be referred to management, make the path followed by each decision as short as possible.

These characteristics look familiar to anyone involved in advanced quality programs. The intent is the same: each individual employee must be accountable for the work he or she is responsible for performing, and each must have the authority to do everything possible to assure the highest quality of that work. These are not the characteristics of a hierarchical structure.

High-Performance Teams

A structure that solves some of the problems of hierarchies is a team approach, known as high-performance teams, among other names. This structure uses a team to perform a process or subprocess that is made up of from as few as 5 to as many as 40 members. They may or may not have a manager in the team. If not, the structure is called a self-managing team, although it will report to a manager who is not a member. One critical aspect of this structure is that the team must receive part of its compensation as a pool to be shared among the members. The amount of the pool depends on the contribution that the team makes to the business, and the success that the business has as a result, but not on the number of staff on the team. There are other nuances to team management, but they vary with the work.

It is the compensation scheme that provides motivation and restricts growth. The team is rewarded for output, but the reward must be shared; so the team is more likely to take on extra work with no increase in staff than the hierarchy. The team is charged with a particular area of work, but they decide among themselves how to do the work and how to assign it. The compensation structure helps in this also: the team will be motivated to assign work to the member who can best do the work for the sake of their common reward. In operation, high-performance teams rely on peer pressure to encourage underachievers, and they apply it readily. These teams are also less likely to allow a nonperformer to remain in a position than is management in hierarchical structures.

The team approaches seem to be getting results that their forerunner, matrix management, did not. The teams run like little businesses within a business. They are used in almost all types of work, from computer operations and restaurants, to professional engineering groups. The companies that have tried this approach include DuPont and Shell Oil, both well established firms.

The Japanese Model and the U.S. Model

There has been considerable criticism of the U.S. management model as compared with the Japanese model, due to the higher level of corporate focus and loyalty that Japanese business have been able to obtain. When these comparisons are made, U.S. business is described as having a competitive basis, encouraging the maximum individual performance, while Japanese business has a cooperative basis, encouraging teamwork. It is unquestionably true that some U.S. businesses have suffered from internal competition and office politics; so there is some acceptance of this comparison, adding credibility to the need for increased teamwork in U.S. industry.

However, it seems to some that the Japanese model may not be optimal, failing to elicit the maximum performance of each member of the company. A more effective model that may be applied to newly forming teams is the one provided by team sports. In sports, the team members are rewarded, indeed retained on the team, only as a result of their individual performance. There is no greater encouragement of individual achievement anywhere. But teamwork is also required. It is obtained primarily by management. The manager of a sport team, from children's schools to professional league sports, is able to identify with great precision the cases of teamwork being ignored to achieve individual goals. The sports coach will first warn and then punish those who place their own interests above those of the team. Sports teams thus achieve the most optimal mixture of individual performance and teamwork through management.

This approach is consistent with the methods of the high-performance team. This analogy may also dispel the impression that high-performance teams are not managed; no one would accuse a sports team of that. The role of coach may be a new one for most managers, however, requiring training in identifying behavior that works against teamwork.

Re-engineering and the Team Approach

Re-engineering is a good time to try team organizations. The techniques of re-engineering do not require that the re-engineered work processes be done by teams, but that the change projects will be run by teams. The transition from hierarchy to team will be less difficult and will present less risk than at any other time.

Several alternatives to team organization will be possible when the re-engineering project enters its implementation phases. The first is to use the change team as the seed team and move the workforce into that team for implementation, but then to implement a traditional hierarchical structure. The workforce, having gone through the implementation as a team, will retain some of the teamwork benefits. The second alternative is to use the implementation team as the working team, retaining the team organization and providing the team with a permanent manager. This will be a team, but the work assignments, motivation, and decisions will be provided by the manager. The third choice is to create the high-performance team, as described, that will require a special compensation scheme, no manager in the team, and the help of a facilitator to get the team started.

In terms of re-engineering, the team approaches all require the same process design work. High-performance teams may, however, require somewhat less individual job definition.

Handling the Difficult Transactions

Any change project can have some side effects that are less than pleasant to deal with. The most difficult, of course, is removing personnel from the workforce, for which the common euphemism is outplacement. In addition, there are problems when any employee loses any form of status. It is tempting to give these problems to the human resources department, but they can do little more than follow the rules laid down for such personnel transactions. The change project itself can do some things, however, both while making plans for the new processes and during implementation.

Outplacement

The prospect of outplacement must be received by the change team with mixed emotions. The fact that there seems to be surplus personnel is the surest sign of success in re-engineering when efficiency is the motivation. It seems essential that some of the staff be removed if the efficiencies are to be realized as actual cost reductions. However, the cost of outplacement can be very great. Experience has shown that staff reductions sometimes do not have their desired effect, as will occur in the following circumstances:

Problems with remaining staff: The remaining staff can react badly when some of their co-workers are let go. Worse, customers can become aware of difficulties in the business and lose confidence in it.

Staff size regrows: The staff levels may increase as cuts reduce the corporate work capability more than expected.

Outplacement expenses high: Outplacement is often more costly than anticipated.

A change project that experiences all these problems will not be viewed as successful, and some have indeed experienced all three. Re-engineering is less likely to result in postchange staff level increases, but it is susceptible to the other two problems. Fortunately, there are methods that can help prevent or control these problems.

The first method is the complete, or nearly complete, avoidance of outplacement. When the size of proposed cuts is very large, for example when a manufacturing site is to be closed, this will, of course, not be possible. Fortunately, re-engineering projects do not normally cut whole sections of the business. The increases in efficiency created by the re-engineering effort can be converted to increased production, increased quality, and even increased marketing efforts. If staff reductions are necessary, some of the reduced effort requirements can be realized by normal staff attrition. Early retirement and other forms of buyout can also be used to reduce another small percentage of the workforce. None of these solutions will solve the whole problem, but all of them together might reduce it from a major problem to a very small one. These approaches are obviously best evaluated by the change team and the human resources department together.

The second method of controlling outplacement is to assure the fairness of the process for those cuts that cannot be avoided. There must be absolutely no use of the change project to settle long-term political disputes, nor to assure the continued employment of cronies. These actions are recognized with amazing clarity by everyone in the company. The damage that unfairness causes is not as clear, but it can be very dangerous. The reaction of staff is that they become self-centered rather than loyal to the company. Their actions after the cuts will be self-serving and defensive; they will place personal objectives far ahead of corporate ones. Many companies have survived mismanaged staff cuts, believing that they have come through without damage because they do not see many resignations. They are mistaken.

Job Grading Problems

The practice of using a rigid job grading system to control compensation is almost universal. Management is nearly forced to use such a system to protect itself from accusations of unfair treatment and not practicing comparable worth. Furthermore, the grading structures must be constructed to prevent long-term employees from getting too much as a result of salary increases. The use of the grading system makes job grades a very important issue to the entire company.

One of the problems commonly experienced in compensation management is grade creep. This results from the pressure put on the system by managers seeking to reward their best performers. Theoretically, performance should not be the motivation for increasing grades. When grade creep has to some extent corrupted a corporation's job grading system, it may affect the re-engineering project in two ways. First, making reassignments will be difficult because the grades are skewed, possibly resulting in the appearance of considerable downgrading. Second, the human resources department may put pressure on the project to deliberately move grades downward.

In a company-wide restructure there is a way to solve this problem to everyone's satisfaction. The entire grading system can be changed. This may give the company the chance to do some other adjustments to the compensation system, such as making parallel advancement paths for managers and individual contributors, such as technicians and salespeople. In some cases it may be possible to use this approach for a re-engineering project by creating a new compensation system for the new processes only, leaving the other parts of the company unchanged. This will give the staff in the new processes an additional bit of status in return for their pioneering. Using new compensation tables and schemes will be especially attractive if high-performance teams will be doing the work in the new business processes.

Reducing the Ranks of Management

Reducing the number of middle managers has become a trend in most large companies. It can be anticipated as a re-engineering objective, mandated by executive management. When this occurs, the questions that the change team will be required to answer will include:

How can layers of management be depicted in the re-engineering diagrams?

How can the number of managers be reduced without radically changing the organizational rationale of the company?

What will be the impact of increased spans of control?

What can be done with a surplus of managers? Should they be outplaced?

To answer the last question first, the extra managers, who are usually the first to be outplaced, should be the last. Unless they are newly hired, middle managers are often the most capable performers in the company with the greatest stores of institutional knowledge. The surplus managers should be used in nonmanagement positions. This can be more easily facilitated if the grading structure allows them a nonmanagement career path that is as attractive as a manager's, and is especially easy if high-performance teams are used. The traditional placement of former managers in staff positions is generally not as good as putting them more directly in the line of fire, such as in change teams or marketing.

The re-engineering charts do not directly indicate the amount of actual work management applied to a process. Approvals and review activities will be shown and presumably minimized by any competent change team. The layers of management, however, show up only in the organization chart. This number will tend to be reduced if the organization charts are drawn from a clean slate based on the Business Activity Maps and Relational Diagrams. To be effective, the change team should be free to suggest adjustments to the management structure where it seems appropriate.

The answer to the questions related to the impact of increasing span of control and reducing the management staff are very important to the success of the project. It is not possible to increase span of control without changing the organizational rationale. It is perhaps oversimplifying to say that, if span of control is to increase, there must be an increase in teamwork, or the work will suffer. However, this statement is for most purposes true. Span of control should not be increased beyond an average of nine or ten without implementing some method to increase teamwork and communication among the staff reporting directly to the manager. Large reductions in management should be balanced by large increases in the reliance on teamwork, unless the management structure was extraordinarily lightly loaded. Some businesses have many positions that are treated as management positions without having management responsibilities; these cases are usually safe for reductions.

Managers are often treated more harshly than other staff during restructuring, because they have fewer legal remedies. However, many of the best employees in the company are often found in this group. They are the richest single source of ideas on change and on gaining a competitive advantage.

Controlling Morale During Re-engineering

It has been said that nobody ever made a nickel from morale. However, it is easy enough to lose many nickels from the lack of it. Companies that do not value the attitudes of their workforce do not value quality.

Change projects of all types have the potential to intimidate and to demoralize the working staff of any company. Changes threaten the security of their positions; in recent years, more often than not, change projects have been thinly disguised staff reductions. However, in most cases it is possible to keep staff morale problems to a tolerable level during re-engineering efforts. In addition to the proper handling of difficult personnel transactions, as discussed, some techniques apply directly to morale itself.

Controlling Morale During the Analytical Phases

Re-engineering has an advantage over other forms of change project in that the effort has more credibility. To actually re-engineer and reimplement work processes, a company will generally be deemed to have goals above and beyond simple staff cuts. This advantage should be exploited during the analytical phases of the project by advertising the project's progress and its findings on a routine basis. Dynamic Business Re-engineering gains some additional credibility from its emphasis on continual re-engineering. It can be said from the onset that the company is changing its practices to include re-engineering, that change will henceforth be a common occurrence and that the company will take special care to see that staff are secure.

Many re-engineering projects issue bulletins periodically. These can be helpful if they do not seem to be concealing or hypocritical. One project team posted Business Activity Maps of the current operation in a hallway so that everyone could see what they were doing.

A very important factor is the treatment that the staff receives from the change team. There must be a certain respect shown for the workforce and for the current work. Those responsible for change sometimes criticize the current process without meaning to offend, but it can offend very much. Despite opinions to the contrary, the average worker cares about the work.

The attention paid to the staff is also important. The famous experiment in lighting conducted in the late 1950s, in which the group with increased lighting and the control group, which received no increased lighting, but who were given as much attention by the experimenters, resulted in both recording equal productivity improvements over the general population in the same facility. It was apparent that the attention itself increased their performance. This applies mostly to the staff that are not directly involved in the re-engineering project.

Measuring Morale Using Re-engineering Tools

The BAM can be used to evaluate the morale factors in the business process itself. The following morale factors may be assessed and indicated on the BAMs:

1. *Staff productivity*: The percentage of value-added time over total time, by staff position and for groups.

2. *Morale factor 1—transparency*: The degree to which an individual staff member can see the whole process, the degree to which it is possible to know what is really going on.

3. *Morale factor 2—personal impact*: The degree of impact the individual staff member has and the flexibility to make their own decisions; the degree to which each staff member can determine the quality of his or her own work.

4. *Morale factor 3—feedback*: The amount of feedback built into the process, which tells the staff member how well he or she is doing, and the speed with which it is available.

5. *Morale factor 4—workload*: The degree to which the workload varies, and the high and low watermarks; workload should not exceed what can be done for long periods, nor should it be too little for optimal morale.

Many other factors contribute to morale, but they are not related to process design.

Controlling Morale During Implementation

It is during implementation that fear begins to grip the workforce in earnest. The most effective morale control instrument during this phase is pure speed. The sooner any adverse effects are behind the project, the better. For example, if the team-building method is used for implementation, those who are not selected for the team should be reassigned very quickly.

Feedback from the project team during implementation is also helpful. This gives the team an opportunity to express itself and to vent any frustrations that are felt. It also lets the rest of the company know what is being done. It is important that the pace of the project be kept very swift during this phase, so that momentum will not be lost, morale will remain high, and, of course, the project will be completed at the earliest possible date.

The most important single support for morale is success. When the project has been implemented and it is successful, the whole company should know about it. This will not only help future re-engineering efforts, it will also boost morale throughout the company. Everyone likes to win.

Building a New Human Resources Capability

To complement and support a continuing re-engineering process, a new human resources management capability will be needed. The corporate personnel support function must be as flexible as the new change paradigm's environment will make the company's business processes. Without this flexibility, personnel actions will always get in the way of needed changes to the business. The new human resources department should be able to take a productive part in the many re-engineering projects that will be undertaken. Organization development expertise is strongly recommended.

The human resources staff is one of the most important participants in the effort; they remove the element of threat from the prospect of change. Their processes and their counseling should help staff to become accustomed to change and to trust the company. Human resources should also find ways to increase the feeling of corporate identity for all employees.

One of the key long-term objectives of using re-engineering is to increase the leverage of individual staff: to increase their output and contribution to the business. The extent to which this can be continued and increased will be the extent to which the company can continue to increase its competitive advantage.

10
Creating a New Business Environment

Working in the change paradigm will create many new practices in a business. The very nature of business may be altered by treating change as an advantage. This chapter discusses the implementation of the change paradigm, and what the world of business will be afterwards.

Change at Work: Increasing Competitive Advantage

The continuing changes that a company will undergo will not be effected merely for the sake of change; they will be undertaken for improvement. The ultimate purpose of change is to increase competitive advantage. With this goal clearly expressed, change can be managed and the participation of the entire company can be focused. The best way for the workforce to look at change is to look through it to the most important objective: winning. The use of re-engineering on a continuing basis depends on establishing this point of view and also on the effective use of change management.

The first consideration in using Dynamic Business Re-engineering is where and how the change management and re-engineering functions fit into the corporation. Without spending more than necessary, the change management function will be required to influence every department in the company, keep track of almost all operational matters, identify problems and opportunities for improvement, and manage all change projects.

Part of the Corporate Culture

The first place that the new change paradigm is implemented is not in any one part of the company; it is everywhere. It is in the corporate culture. Convincing everyone in the company that change is not a threat immediately becomes one of the greatest advantages of the new approach. When change is no longer feared, it will become a focus of corporate efforts, a galvanizing force. Change will be seen as an opportunity to control the fate of the company. The energy spent defending against change can then be channeled into improving the business. Ideas for improvement will come from all levels in the organization.

There are essentially three steps in making the change paradigm part of the corporate culture. In the initial corporate Positioning effort, senior management should:

1. Ask for the trust of the workforce.
2. Conduct the Positioning/re-engineering process openly.
3. Make special policy changes to protect the workforce, such as a new job grading system that allows staff to be moved freely (see Chap. 9).

Second, and most important, all promises are kept during the first actual re-engineering effort and hopefully all subsequent ones. If any staff cuts are to be made, they should be done at once and kept visibly separated from the re-engineering effort. Third, the Positioning infrastructure, which includes the permanent re-engineering group, the tools, and the baseline business data, are made visible and accessible to the entire workforce. Every member of the company is thereby encouraged to trust in the change processes and participate enthusiastically in them. A sense of ownership in the change process will develop in time; staff will feel confident about it.

There Is a Price

These new capabilities are not free. Some initial investment must be made, and there is also a price for maintaining the change machinery in working order. The cost and organizational factors related to these efforts should be kept small, not only to conserve costs, but to prevent them from establishing their own bureaucracies.

The first requirement is for a new executive position to manage the change process, the Positioning effort and the re-engineering projects: the chief change officer. It is recommended that this position report to the chief executive or chief operating officer, so that it will be included in high-level interactions and also so that it will not develop an imbalanced orientation toward one aspect of the business. For example, if the position reports to the chief financial officer, it may lose its operational orientation. The chief change officer will also require a grasp of the technology used in business (see "Computer Management Literacy," page 195).

The second requirement is for a permanent re-engineering group, the Positioning team (see Chap. 6), responsible for the initial gathering and maintenance of the corporate baseline data and the re-engineering tools. The speed with which re-engineering can be done depends on the success of these efforts. The members of this group also form a pool of re-engineering experts that provides members for the re-engineering change team of each new project. They should have varied backgrounds in analytical fields, and some of them should have long experience with the company.

The third and last infrastructural cost is for the tools themselves. There will be initial costs and maintenance, possibly upgrades of new versions of the various technology tools, and perhaps new re-engineering tools as more become available. Most of the tools will be personal computers and related software. The costs will be small, especially compared to the support they provide and the potential payoff of re-engineering itself.

Change Management and Its Peers

The installation of change management in a company should require neither a large organization nor a great expense. However, it should result in a new function that will be as important to the company as its other major functions. By keeping the costs and size of the re-engineering group as low as possible, it may be difficult to justify the group's peer position with such as accounting or even business operations, but the entire change management functions depends on this relationship.

Another factor critical to the viability of change management is that it be used. If long periods go by either without change projects, or between them, each new project will be a start up effort and the benefits of continuing change will be lost.

This function is new to business and the use of an internal group to manage it is even newer. So far, most re-engineering has been done with considerable guidance from consultants. However, an environment of change cannot be managed on a long-term basis from outside the company. Furthermore, it is difficult to obtain continuing competitive advantage from consulting alone, since consulting is a service for sale to everyone. Therefore the establishment of an internal corporate re-engineering group is a cornerstone of the change paradigm.

Using the Tools Regularly

The first corporate re-engineering project will do more than break new ground. All of the tools (such as, computer systems) and techniques required for continuing change will be acquired and put into use for this first effort. To reuse the tools, certainly if the Dynamic Business Re-engineering methodology is used, it is necessary only to leave them in place at the end of the first project and establish the infrastructure as described.

To Change and Tune the Business

The principal use of the models and other change information maintained by the Positioning team will be to support future re-engineering efforts. Some of the projects will make major changes, others minor tune-ups. In some cases, however, business operations can be tuned with the help of this change information and the company's re-engineering tools, but without re-engineering projects.

The models will keep track of many of the businesses work parameters. For example, the length of time each process takes is stored in the database. If a process can be improved, the models will demonstrate the impact, even if no change in work flow is needed. If a process time can be reduced by a simple change in the quality of a supply item, then no re-engineering need be done. Management will, however, want to know the results and compare them with the cost of the change.

The use of the models for these minor efforts serves to keep the data up-to-date, to make staff more familiar with the change process and to provide valuable decision support.

To Modify and Enhance Information Systems

Dynamic Business Re-engineering methods are used to define information systems and may also be used to maintain them. Even when the information systems were not generated by the re-engineering tools, the tools can be useful in locating requirements for enhancements to the systems.

When old systems are being used, re-engineering and CASE tools may provide entirely new systems at a lower cost than would be required to maintain and enhance the existing systems for only a few years. Maintenance costs remain very high for systems built by the best technologies available a few years ago. Systems older than a few years may be nearly impossible to maintain. The combination of re-engineering and CASE tools can construct computer support for a fraction of the costs of the older technologies, and they can be maintained more cheaply. This capability will put applications systems development on a renewed footing in business, and will enable many unfulfilled demands for computer support to be met.

To Manage the Operation

The tools and models of re-engineering will become valuable to management. They will provide operations documentation that is lacking in most businesses, and performance standards that can be used to measure efficiency.

Innovative managers will find other uses for the Positioning models if they are allowed access to them. For example, the process models could be used to track the production of individual orders as the work steps weave their way through the company.

The process models can also be used to train both management and staff. They can support cross-training of personnel for contingency support of alternative processes and make the transfer of staff easier. The same could be done for the company's ability to rotate managers.

Of course, the daily use of re-engineering tools increases their effectiveness in change projects and increases the trust of the company in their capabilities.

Controlling Change

Change must obviously be controlled. However, when an environment of continual change is adopted, managing change itself changes. The goals of change management are the avoidance of confusion and achievement of objectives. Continual change presents special problems in both areas. Fortunately, these problems can be controlled by the Positioning and re-engineering methods themselves.

Coordinating Change Information

A change-friendly environment implies ready access to change information: the information that Positioning gathers and maintains. The change management group gathers this information, maintains it, and makes it available to everyone in the company. How is this done?

First, the initial Positioning effort starts the process. The information is gathered by the techniques described in Chaps. 6 and 7. This data is analyzed, organized, and entered into automated databases. When the re-engineering process starts, a portion of the data is turned over to the re-engineering change team. The Positioning baseline will remain available to the company for reference during the re-engineering project, but only the project team will be able to change data related to the work that they are doing. As the project team develops new models of the operation, they are given project version numbers and indexed.

At the end of the re-engineering project, a final check is made of the project by the Positioning team. The checked design becomes the "as built" process maps, and are returned to the control of the change management group with new production version numbers. The effective dates of the new designs are also noted.

Changes can be expected to this data on routine basis when the processes are in production. These changes are gathered by having all change transactions pass through the change team. Most of this activity can be automated. To prevent conflicts, these smaller ongoing changes are controlled by the change team and not the permanent change management group. Routine changes are always dated as they are entered. In this way, the change information databases provide a snapshot of the company and constitute an audit trail.

The overall quality of the data is controlled upon entry and is checked by periodically checking the models against real processes and primary data sources. This provides high-quality data bases and reliable reports.

Manual and Automated Models

There is nothing in re-engineering that cannot be done with paper and pencil. Even the Dynamic Business Re-engineering methods do not necessarily require automation. However, as continuing advances are made, it becomes increasingly beneficial to use some automated support. Using paper models makes extensive reworking of processes difficult, which may limit the quality of the design work. It is also very unlikely that any organization will keep data stored on paper up to date, making reuse unfeasible and, in essence, making the change paradigm itself untenable.

The automation of re-engineering is at an early stage today. Other than the PAR system, no computer software products support change management and re-engineering design. And even PAR does not have its own internal project management software, nor does it currently directly generate computer systems to support new business processes. So the change management groups are currently forced to use more than one system. Fortunately, the systems available work well together.

The most basic automated tool is the word processor. Because some of the re-engineering work will require computer graphics, a graphical user interface (GUI) word processor is the best choice. These include any word processors running under the following: Microsoft Windows, IBM OS/2, Next, and Apple Macintosh. The drawing of process flows, if BAMing or any other graphical methods are used, can be done by personal computer drawing systems, CASE tools, or the PAR system. Drawing tools do not have the ability to redraw flows maintaining connections, so they are the least attractive choice. CASE tools, intended to aid the design of computer systems, are a better choice, since a program flow or data flow diagram has many of the characteristics of a work flow chart or BAM. CASE tools cost much more than drawing software, but they are worth the extra price to the re-engineering project team. Today's CASE tools will not store much of the change data, and will not associate that data with the process charts. An important facility that many CASE tools provide is distributed access; they can make their data available over local networks or on mainframe computers for wider use.

Re-engineering projects, especially large ones, may want to use project management software to support planning, resource allocation, and monitoring. Several project management software packages are available, such as Microsoft Project, Lotus Agenda, and Symantec Timeline. An additional requirement for automation is the generation of applications software systems to support the new processes designed by re-engineering. This capability is also rendered by some of the CASE tools, although the ability is currently somewhat limited. If systems work will be performed in support of re-engineering, the Information Services department should have the same CASE tools as those being used by the change team.

Batching Changes

As desirable as the capability to continually change the business may be, making substantial changes on an hourly or daily basis is not feasible. If there is no time between changes in a given process, there will be no chance to evaluate the change, no time to learn how to operate in new environments, and no work done. The company will spend all its time and resources making changes. Also, making very small changes on a very frequent basis is difficult to control and the changes may be hard to back out of production if they do not work well enough.

To avoid the problems associated with constant uncontrolled change, the small changes that occur frequently can be accumulated into batches of changes, and made at predetermined times. Each major process should have its own batches, and they should be staggered company-wide, so that the whole company is not changing at the same time, unless there is an operational reason to do so.

The change batches will each have impacts and repercussions beyond the processes being changed, and the changes to processes will cross organizational lines. As a result, re-engineering techniques will be needed to coordinate the implementation of the change batches (see Chap. 7).

Managing Politics

Politics and internal competition are present in all companies. Managing politics—that is, redirecting the attention and energy spent on infighting to productive use—is one of the most difficult problems facing established businesses today. However, solving this problem is essential to successful re-engineering. The results of negative political action can destroy any re-engineering effort.

The Hidden Agendas

The principal consequence of politics and self-interest is the "hidden agenda": the private plan of the corporate plotter. These plans quickly influence re-engineering, as the managers and possibly team members see opportunities for themselves and begin to subvert the project to their own purposes. Also, conflicts arise when hidden agendas clash. As the agenda holders are not able to support their cases with their real motives, complex rationalizations are often devised. Resolving these problems can cause significant delays and injuries to the interests of the company.

Unfortunately, the managers who display this behavior cannot be excluded from the re-engineering project without losing the benefit of their knowledge and their commitment to the changes being undertaken. The only remedy is fast, focused action by senior management.

Controlling Politics from the Top

Only the highest levels of management in the company have the leverage and the abilities required to control politics. The best approach is the one used by sports coaches:

1. Emphasize and practice teamwork.

2. Recognize and reward teamwork.

3. Recognize and immediately punish behavior that is contrary to teamwork.

The heads of most companies know what political behavior looks like. Effective responses can take many forms, from a quiet talk to public censure. For example, a well known "un-teamwork" practice is to complain about a peer to higher-level management without attempting to resolve the problem first with the peer. One executive made a consistent practice of rerouting memos that contained complaints to the manager about whom the complaint was made, which usually sent the right message to the sender as well. The pressure must be kept up and must be practiced by all the top executives for a real change to be made, especially in companies where politics is part of the culture.

Most senior executives know who causes the most damage by pursuing self-interest. In some cases this behavior is excused as being a natural consequence of a high level of motivation, and it is accepted because it accompanies high levels of performance. There are senior managers who consider internal competition to be beneficial. They believe that the competition is under control and that the result is increased performance. However, if more control were to be exercised, the politics would decrease, but the performance would not. The same people would be involved, and these energetic, ambitious, perhaps ruthless managers would continue to put forth their best efforts to succeed.

Start-Ups Using Dynamic Re-engineering

Re-engineering methods can also be used for start-up projects where there are no existing processes to serve as a baseline. While these cases may seem simple, there are a few special concerns.

Starting a New Operation

New business processes are not complicated by old investments and old ways of doing things, which gives them certain advantages. In fact, one of the objectives of re-engineering is to give the established businesses the advantages of new ones. However, the advantages attributed to a new business are generally made in reference to new businesses that have succeeded, ignoring the ones that have failed. Out of at all business attempts, a high percentage of companies do not survive their first two years. A common cause of these failures is cited as undercapitalization, which is not something re-engineering can help. Re-engineering can, however, help with a wide range of other problems that are related to business process issues.

One problem that new businesses and new processes in established businesses must face is that they do not start from a solid basis of known feasibility. It is possible to design a new process that will simply not work. The re-engineering of old processes is protected from this to a large degree by having the performance and requirements data from a functioning process as a starting point. There are a number of ways to help assure feasibility:

1. Some members of the design team should be experienced in the business processes being designed.

2. Studies published about other companies' experience can be consulted.

3. Professional consultants with experience in the processes can be engaged.

But the most effective approach is to use the continuing change process. The final implemented design will be expected to produce results that will be documented and measured. If the process does not produce the expected results, it can be re-engineered and reimplemented, correcting any problems that are observed.

Creating Models from Scratch

Creating Business Activity Maps or work flow models from a clean slate should not present any special problems. Like the ones produced from existing processes, they are developed from the top down by breaking general tasks into greater levels of detail. However, when designing without a real process, the lowest levels of detail may not reflect reasonable job specifications for the staff who are to do the work. To reduce the risk of designing an unimplementable job, it may be desirable to break down the details, try to assemble an organization chart, and then redraw the detailed activity maps based on what has been discovered.

Again, team building and continual re-engineering can be used to create a trial process, and then tune it based on results. A flexible team should have no trouble in accepting the ongoing reassessment and modification of their own work processes. The natural stimulus and optimism associated with start-up efforts should provide a high level of cooperation and motivation for this effort.

A Vision of the Future

The prospect that the future holds more change than the recent past can be unsettling. If the future will be so different that no predictions can be expected to present even a reasonable risk, then how can any plans be made? How can any preparations be expected to give advantage? Is the safest course just to invest in gold and wait?

The answer is that, even with the pace of change increasing, the direction of future events is not completely unpredictable. From the business viewpoint, there are only three major scenarios:

1. The complete annihilation of humanity (the pessimistic scenario).

2. A great change in technology, but little in society (the status quo).

3. A great change in society for the better (the optimistic scenario).

The pessimistic scenario seems to have vanished with the end of the cold war, and the status quo is definitely changing. The future will most likely be somewhere between the status quo and a great improvement in society. Some of the character of this improvement will involve business and business can take advantage of it. Those preparing themselves will not be wasting their effort.

A New Marketplace, a New World

All businesses are moving into a new world, whether they are prepared or not. This new age is not confined to business, but it seems from the events of the recent past that business is about to increase in influence. All the military activity during the cold war failed to decide the issues of that age; economics and commerce finally overwhelmed force, ideology, and politics. We are now in a world in which there is still conflict, to be sure, but it seems to be sublimated to the universal desire to achieve, to improve, to excel: in a word, to win. This gives humanity an opportunity that has never presented itself in all of the 6000-year history of civilization. From the current political and social climate, there can grow the first great global civilization.

Great civilizations of the past have been localized. Egypt, Greece, Rome, China, the British Empire, and some would include the United States have been countries in a larger world. The reasons for their greatness have been a source of speculation and study, but it is quite probable that it was the environment and not individual people behind the growth of each of the world's great cultures. The worldwide environment today may contain the basic ingredients for a great civilization. First, there is no great polarizing conflict to frustrate international cooperation and commerce. Second, there is growth: there are now serious efforts being made by most countries to develop themselves. The spectacular rise of Japan, Singapore, and South Korea has demonstrated that it is possible for effective government and social action to create commercial empires from small and war-ravaged countries. The role of government in these countries seems to be supporting globalized private business in a new and more effective way. Government ownership and government control are not part of these success stories, but cooperation and encouragement are. These countries have provided a glimpse of the future. Although their specific methods may not be transferable to other societies, their principles will be. One of these principles is that competition in business is a productive basis for social interaction.

Re-engineering: The Fundamental Tool of Change

Re-engineering is both the fundamental and ultimate tool of change. Re-engineering addresses the business process, which is the means by which work is done, whether by people or machines. In its present state it is helping to adjust business from the old industrial paradigm to a new one of service and information. In the future, it will continue to move business from one paradigm to the next.

As re-engineering is used, it will itself undergo several paradigm shifts. Business is going through one now, but there is at least one more on the horizon: the second paradigm shift—using continuing change for competitive advantage.

Business opportunities will continue to grow. One way or another, the relationship between government and industry will improve everywhere. Connections between business, education, science, technology, government, labor, and financial services will become increasingly supportive of success in enterprise. Most of the benefit of these improvements will come to business without much effort. However, the businesses that gain the most will be those that can assimilate the most new technology and take advantage of opportunities with the least delay. They will be the businesses that equip themselves to change.

Index

245

About the Authors

Daniel Morris and Joel Brandon are principals of Morris, Tokarski, Brandon & Co., a Chicago-area-based management consulting firm specializing in business positioning and re-engineering. In the vanguard of the re-engineering movement since the concept emerged, they each have more than 20 years of executive and management experience in a wide range of industries. Their consulting activities have included operational re-engineering, strategic and tactical planning, operational evaluation, project management, and government policy making.

Both authors have spoken widely on a variety of business topics, and each has written many of the seminal articles on re-engineering. Their previous collaborations include a series of video training courses for Applied Learning, Inc., and the Information Continuum/Information Management Series, as well as their book *Relational Systems Development* (McGraw-Hill), which introduced a method for relating information technology support to the business operation.